T0203796

Principles of Population Dynamics and Their Application

Alan A. Berryman
Washington State University

Taylor & Francis
Taylor & Francis Group

LONDON AND NEW YORK

First published 1999 by: Taylor & Francis
2 Park Square, Milton Park, Abingdon, Oxon, OX14 4RN
270 Madison Ave, New York NY 10016

Transferred to Digital Printing 2008

A catalogue record for this book is available from the British Library

ISBN 0-7487-4015-5

Typeset by The Florence Group, Stoodleigh, Devon

Publisher's Note
The publisher has gone to great lengths to ensure the quality of this reprint but points out that some imperfections in the original may be apparent.

Dedication

For Annie Augusta

Dedication

Contents

Preface

If there is one thing that can be said for the universe, it is that all things change with time. Nothing is still. Everything is in motion. And as for us? We can try to understand and predict the inevitable changes, or even to alter them to our own ends, but we cannot avoid them.

This book is about ecological *change* and, in particular, about changes in populations of living organisms. In other words the book is concerned with the *dynamics* of *populations*.

This book is about *detection*. Understanding why a particular population changes in a particular way is much like uncovering the person who committed a particular crime. Like a detective, we try to deduce the underlying causes of fluctuations in natural populations from clues in the empirical evidence, and from our understanding and intuition about the basic processes involved.

This book is about *diagnosis*. Certain diagnostic procedures will be used to search for clues in the ecological data, much like a physician uses X-rays, blood samples, electrocardiograms, and so on, to search for clues to your state of health. Professional ecologists have much in common with physicians and detectives, for they all have to deal with unique and extremely complicated systems that are almost impossible to understand in detail, they all use diagnostic tools to look for clues or symptoms, and they all develop theories or hypotheses about the causes of the problems they are investigating. The ecologist is both detective and physician – surely an exciting profession!

This book is about *forecasting* ecological changes. Ecology has been criticized for its overemphasis on retrospection, or how we got to be where we are, and lack of concern with prediction, or where we are going in the future[1]. Evolutionary ecology, about which much has been written, is mainly retrospective. In this book we walk the predictive path because we are mainly interested in answering the question: Given that we have a certain set of conditions at *present*, what changes can we expect in the near or distant *future* and how can we affect those changes to our own ends, or to the ends of the planet? This is not to say that retrospection or evolution is unimportant, for it may be necessary to understand how we got to be where we are in order to see where we are going. This book, however, emphasizes the predictive rather than retrospective side of ecology.

This book is necessarily formal and *mathematical*. Models are needed to forecast ecological change. Mathematical models are preferable, for then others can make unambiguous forecasts with the same instrument. It is important, therefore, that the student has a basic training in mathematics – at least college-level algebra and statistics.

This book employs a blend of *empiricism* and *theory*. There are several different ways to model natural phenomena[2]. Some people like to include as much detail as possible, even going so far as to model each individual as a separate genetic entity. These highly detailed and realistic models are usually called simulation models because they can only be solved by simulation on a computer. Others prefer a purely statistical approach, modelling the object of interest as a function of one or more independent predictor variables using statistical techniques. There are problems with both of these approaches, however: simulation models may provide realistic descriptions of the system but are often incomplete because of data shortages, and may be inaccurate because errors tend to be amplified by the complex simulation process. Empirical models, on the other hand, may be quite accurate, but they provide little insight into the underlying biological processes because the model is statistical rather than ecological. The approach taken in this book is partly empirical, in that statistical procedures are used to search for clues in real data (diagnosis), and to fit models to those data, but the underlying models are based on ecological rather than statistical theory. In other words, the approach we will use combines predictive empiricism with explanatory theory.

This book emphasizes the *practical* side of ecology. It is aimed at applied ecologists, resource managers, and pest managers. It is also aimed at the undergraduate student taking courses in pest, fisheries, and wildlife management. Unfortunately, ecology is cursed with confused terminology and theoretical controversies. Together with the mathematical demands of a predictive science, this has made population dynamics the dread of undergraduate students in forestry, fisheries, wildlife and pest management. I have done my best to reduce the student's dread by paring away or avoiding much of the controversy, complication, and jargon. In other words, I have tried to reduce population dynamics to its bare essentials. Some experts may find this approach oversimplified, while others may be dismayed by my dismissal of their favourite terms and/or theories. But this book is not written for experts. I can only hope that my reductionist approach will appeal to teachers and students of applied ecology.

In order to keep the book simple and to avoid confusion, I have tried to keep the main text free of literature citations and terminological or theoretical arguments. When citations and detailed explanations are considered necessary, they are marked by superscript numbers that identify their place in the **Reference** section at the end of the book. The text is written with the advanced undergraduate student in mind, but the references are supplied to stimulate the thinking of graduate students and professionals. I hope my colleagues will forgive my liberties with the literature and their ideas.

This book contains data from real resource and pest populations. Ideally, this book should be used in conjunction with laboratory sessions during which students can practise analytical and computational skills. For these purposes, a series of microcomputer programs have been developed for performing the analyses presented in this book. These programs are known as the *Population Analysis System*, or *PAS* for short, and can be purchased from *Ecological Systems Analysis*, 921 West Main Street, Pullman, Washington 99163, USA. Although these specialized programs may simplify the problem of analysing population data, they are in no way essential. All the techniques used in this book can be found in standard statistical software packages and most can be performed on standard microcomputer spreadsheets.

Finally, this book represents the combined ideas of many people, a few of which deserve special mention. My students have contributed most as their young minds challenged the conventions of contemporary science. The development of the microcomputer programs, now called the *Population Analysis System* or *PAS* for short, was initiated by Jeff Millstein while he was my graduate student in the mid-1980s. It was he that dragged this old "main-framer" into the new age of microcomputers. My general understanding of population dynamics was greatly influenced by Russian forest entomologists and mathematicians working under the direction of Academician Alexander Isaev, and by the rigorous but difficult monographs of Tom Royama. Recent forays into predator–prey theory, a necessary ingredient for the completion of this work, were aided by my interaction with Roger Arditi, Lev Ginzburg, Jurek Michalski and Andrew Gutierrez. The diagnostic approach developed in this book has benefited greatly by interaction with Peter Turchin. Although our approaches are somewhat different, our objectives – the analysis and understanding of population dynamics – are the same. I would also like to thank Brad Hawkins, Peter Price, Peter Turchin and an unknown reviewer for reading this book in its entirety before it went to press, and Bill Mattson, Doug Johnson, Mikael Münster-Swendsen, Amos Barkai, Erik Christiansen and Monique Borgerhoff Mulder for reading the chapters on cone beetles, sandhill cranes, spruce needleminers, rock lobsters and human beings. Their comments, constructive criticisms, and unpublished data were invaluable in the production of this work.

The following are acknowledged as sources for the quotes beginning each chapter and elsewhere:

Allee, W. C. (1931) *Animal Aggregations: a Study in General Sociology*. University of Chicago Press, Chicago (Chapter 4).

Andrewartha, H. G. and Birch, L. C. (1984) *The Ecological Web*. University of Chicago Press, Chicago (Chapter 2).

Barkai, A. and McQuaid, C. (1988) Predator–prey role reversal in a marine benthic ecosystem. *Science* 242, 62–64 (Chapter 15).

Blackman, M. W. (1931) The Black Hills beetle (*Dendroctonus ponderosae* Hopk.). Bulletin of the New York State College of Forestry 4, 1–97 (Chapter 14).

Darwin, C. (1859) *The Origin of Species by Means of Natural Selection*. Murray, London (Chapter 5).

Gause, G. F. (1934) *The Struggle for Existence*. Williams and Wilkins, Baltimore (Chapter 10).

Hutchinson, G. E. (1948) Circular causal systems in ecology. *Proceedings of the New York Academy of Sciences* 50, 221–246 (Chapter 6).

Ivlev, V. S. (1955) *Experimental Ecology of the Feeding of Fishes*. Yale University Press, New Haven (Reference 10).

Johnson, D. H. (1979) Lesser and Canadian sandhill crane populations, age structure, and harvest. *United States Department of the Interior, Fish and Wildlife Service, Special Scientific Report, Wildlife No.* 221 (Chapter 11).

Keynes, J. M. (1951) *Essays in Biography*. Granada Publishing, London (Chapter 16).

Malthus, T. R. (1798) *An Essay on the Principle of Population as it Affects the Future of Society*. J. Johnson, London (Chapter 3).

Maruyama, M. (1968) Mutual causality in general systems, in: J. H. Milsum (Ed.) *Positive Feedback: a General Systems Approach to Positive/Negative Feedback and Mutual Causality*. Pergamon Press, Oxford (Reference 21).

Mattson, W. J. (1980) Cone resources and the ecology of the red pine cone beetle, *Conophthorus resinosae* (Coleoptera: Scolytidae). *Annals of the Entomological Society of America* 73, 390–396 (Chapter 12).

Munster-Swendsen, M. (1985) A simulation study of primary-, clepto- and hyperparasitism in *Epinotia tedella* (Cl.) (Lepidoptera: Tortricidae). *Journal of Animal Ecology* 54, 683–695 (Chapter 13).

Newton, I. (1729) In *Newton's Principia: Motte's Translation Revised*, by F. Cajori. University of California Press, Berkeley (1960) (Chapter 7).

Royama, T. (1992) *Analytical Population Dynamics*. Chapman and Hall, London (Chapter 9).

Simpson, G. G. (1961) *Principles of Animal Taxonomy*. Columbia University Press, New York (Chapter 8).

Slodbodkin, L. B. (1968) Animal populations and ecologies, in: J. H. Milsum (Ed.) *Positive Feedback: a General Systems Approach to Positive/Negative Feedback and Mutual Causality*. Pergamon Press, Oxford (Chapter 1).

Finally, the picture on the cover of this book was made from a photograph of a ladybird beetle aggregation taken by Professor Roger D. Akre (deceased) while he was a faculty member in the Department of Entomology at Washington State University.

AAB
Pullman, 1997

PART ONE

Theory

Dynamics: theoretical ecology and the rules of change | 1

Populations of organisms can meaningfully be analyzed as feedback systems of a rather complex kind, showing both positive and negative feedback loops. (L. B. Slobodkin, 1968)

Numbers of organisms change in interesting and mysterious ways (Figure 1.1). Some, like the human population, seem to be growing continuously, while others, like the blue whale, seem to be on their way to extinction. Growth patterns such as these have important implications for the species in question, for those utilizing them as resources, and for the persistence of life on this planet. Obviously, the human population cannot grow forever, but when will it stop growing and what will stop it? Are blue whales becoming extinct and, if so, what should be done to aid their recovery?

Other species, like the sycamore aphid in England, appear to remain remarkably constant for long periods of time, even though they may fluctuate considerably from one year to the next. What keeps these populations from increasing like humans or decreasing like whales, and why do they exhibit the sharp, short, 2-year fluctuations called, in this book, high-frequency or "saw-toothed" oscillations?

Yet other populations, like the larch budmoth in the Swiss Alps, go through dramatic and very regular multi-annual oscillations which often take around 10 years to repeat themselves. Although ecologists have been fascinated by these so-called "population cycles" for more than 50 years, there is still no general agreement about their causes.

Other species have much more variable dynamics. Gypsy moth egg mass counts in New England, for example, declined suddenly and, at the same time, switched from a saw-toothed to a cyclical pattern of fluctuation, while whitefish in Lake Ontario remained relatively constant for a long time but recently entered a period of decline.

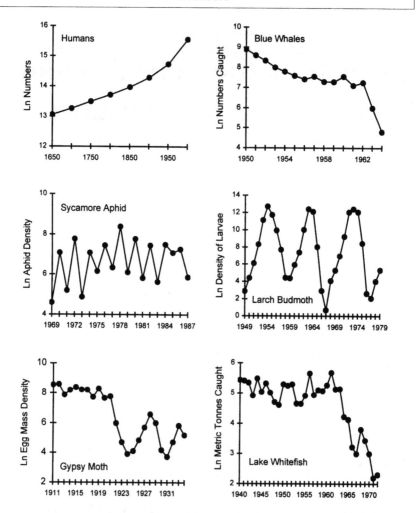

Figure 1.1 Dynamics of various animal populations with numbers, expressed as natural logarithms, plotted against time (see reference 3 for data sources).

1.1 THEORETICAL ECOLOGY

Why do some populations of living organisms grow and decline, while others oscillate around a relatively constant average density? Why do some oscillate rapidly while others cycle slowly? Theoretical population ecology attempts to answer these and other similar questions, to make sense out of the complexity and confusion we see in nature, to uncover the general rules or principles that all populations of living organisms must obey.

Students sometimes ask why they have to learn abstract theoretical concepts when what we are really interested in is solving real practical problems, such as how to harvest renewable resources, reduce the ravages of pests, or save endangered species. The best answer to this question may well lie in a quote borrowed from a forgotten writer: *"theory without fact is fantasy, fact without theory is chaos"*. What this statement means is that theories should be based on factual information and that, once established, the theory should be used as a framework, or a set of guidelines, for organizing facts and actions. A good example is the space programme: although it was modern engineers who designed and built the vehicle that put men on the Moon, it was the theories of gravity and planetary motion, laid down more than 300 years earlier by Kepler and Newton, that showed how it could be done. Without theory, putting a human on the Moon would have been impossible.

Resource and pest managers sometimes seem reluctant to pay the same attention to theory as space engineers, and have occasionally reaped the dire consequences. A good example is the use (or misuse) of the chemical DDT to control insects. After its rediscovery in 1939, DDT was hailed as a miracle insecticide and was widely used against all kinds of insects without regard to the theoretical consequences of such actions. Those scientists who warned that insects would become resistant to the chemical, as predicted by the theory of evolution, were ignored or ridiculed. Yet they were right, for DDT became more or less useless against the many insect pests that evolved resistance to it. Theory, therefore, should be considered an essential ingredient of pest and resource management policy[2].

A theory can be defined as a systematic statement of the principles, processes and relationships that underlie a particular natural phenomenon. Thus, theories attempt to explain observed events by reference to known principles, relationships and causal processes. For example, the theory of evolution attempts to explain the complexity and diversity of life on this planet from the basic principles of heredity, variability, and natural selection. Similarly, population theory attempts to explain the complexity and diversity we observe in the fluctuations of natural populations, such as those illustrated in Figure 1.1, by appeal to the basic principles of dynamic systems and to the ecological processes that evoke those principles. Thus, in order to develop a theory of population dynamics, it is first necessary to understand something about the general rules of change, rules that apply to all dynamic systems; automobiles, rocket ships, television sets and ecosystems.

1.2 THE RULES OF CHANGE

Changes in a dynamic system can be caused by either *exogenous* or *endogenous* processes. For example, the predator population in the top diagram of Figure 1.2 acts as an exogenous effect on the prey population

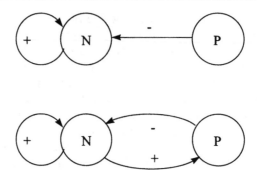

Figure 1.2 Flow diagram for the effects of a population of predators P on the numbers of its prey N: top, predators act as a negative exogenous factor on the prey population (i.e. negative arrow from P to N), and the prey have a positive effect on their own numbers (i.e. the arrow from the prey to itself has a positive sign); bottom, prey now have a positive effect on the predators so that the interaction forms an endogenous feedback loop [Note that the feedback is negative because the product of the signs in the loop is negative; i.e. $(+) \times (-) = (-)$].

because it causes changes in prey numbers without being affected itself by those changes. This is shown as an arrow with a negative sign going from the predator to the prey population (ignore the positive arrow from the prey to itself for the moment). The arrow is negative because changes in the prey population are inversely related to the number of predators; for example, an increase in the number of predators causes a reduction in the prey population because more are eaten. Notice that the prey population has no influence on the predator (no arrow from prey to predator). In this sense, the predator population is external to and independent of prey numbers and, because of this, is considered to be an *exogenous* influence on prey dynamics.

Endogenous effects, on the other hand, cause changes in a dynamic variable and are also affected in return by those changes. For example, in the bottom diagram of Figure 1.2, the prey population is now assumed to have a positive effect on the predator population (e.g. an increase in prey numbers leads to more predators because, given more food, they produce more offspring). The addition of this arrow links the prey and predator populations together in a mutually causal *feedback loop*, and the two populations are now considered to be parts of the same closed system. In this sense, predators are now an *endogenous* influence on prey dynamics.

1.2.1 Endogenous dynamics

Feedback loops are created whenever a variable influences itself, either directly or through another variable (Figure 1.2). Feedback loops are identified by the sign of the feedback (+ or –) and by the number of components that are involved in the loop (*order* or *dimension*). For example, a *first-order* feedback loop possesses only one component (e.g. the positive loop from the prey population to itself in Figure 1.2), while a *second-order* feedback loop possesses two components (e.g. the feedback loop involving both prey and predator populations in Figure 1.2, bottom). As we will see below, the sign and dimension of the feedback have important effects on the *stability* of dynamic systems.

Dynamic systems are said to be *stable* if their variables return to, or towards, their original states following disturbances. For example, the temperature of a room is stable if, when a door is opened (= exogenous effect), the thermostatic control system brings the room back to, or at least towards, its previous temperature (= endogenous process). This does not mean that the temperature will be exactly the same as that set by the thermostat, but it will not deviate very far from that setting. Here the thermostat set point is also called an *equilibrium point*. Hence, stable systems tend to remain in the vicinity of an equilibrium point and to persist in a state of balance, even though the environments in which they exist may be quite variable. They are said to be *homeostatic* or *regulated*. Notice that, although an equilibrium or set point always exists in a stable dynamic system, this point may never actually be observed if the system is subjected to continuous exogenous disturbance.

Stability in dynamic systems is brought about by the action of *negative feedback* processes (in shorthand notation ¯feedback). Hence, it is ¯feedback that acts to oppose changes in the state of a dynamic variable, just as the thermostat opposes changes in room temperature. The thermostat is, in fact, part of a mechanical ¯feedback mechanism. Natural ecosystems can also create ¯feedback processes. For example, consider the second-order feedback loop created by the interaction between populations of predators and prey (Figure 1.2, bottom). Here an increase in the number of prey leads to more predators and this larger population of predators then reduces the prey population back towards its previous level of abundance. As a result, the prey population tends to be stabilized or regulated by the interaction with its predators. Notice that the sign of the feedback loop is determined by multiplying the signs of the individual interactions within the loop. In the case of the predator–prey interaction we have a positive effect of the prey population on the predator population and a negative effect of the predator on the prey, or $(+) \times (-) = (-)$.

Although ¯feedback is a *necessary* condition for stability in dynamic systems, it is not a *sufficient* condition, for in order to have a high degree of temporal stability, the ⁻feedback must also occur rapidly. It is fairly obvious that the operation of a second-order ⁻feedback loop, such as that in Figure 1.2 (bottom), must take some time, because time is needed for the predators to react to an increase in their food supply and to turn that food into offspring. This is called the time *delay* or time *lag* in the feedback loop. In general the greater the order or dimension of the feedback loop, the longer the time delay. Time delays in ⁻feedback loops tend to introduce oscillatory instability into the system, with longer time delays causing oscillations of longer period and greater amplitude. For example, one explanation for the 10-year cycles observed in larch budmoth populations (Figure 1.1) is that defoliation of larch in 1 year reduces the quality of the foliage in the following year. As a result, the reproductive rate of the insect in that year is affected by the density of the population in the previous year, giving rise to a time delay of 1 year in the ⁻feedback loop. Note that the ⁻feedback between insect and foliage has a dimension of two (or second order) because two components are linked together in a mutually causal feedback loop (e.g. in a similar way to Figure 1.2, bottom). In general then, the higher the order or dimension of the ⁻feedback loop, the longer it takes for the effects to be transmitted through the loop, and the less stable the dynamics of the system. For this reason, high-dimensional ⁻feedback processes are likely to be involved in many of the periodic rhythms observed in nature.

Dynamic systems are said to be *unstable* if their variables continue to move away from their original states following an exogenous disturbance. The human population, for instance, is currently exhibiting unstable dynamics because it is continuously increasing. The blue whale population may also be unstable because it seems to be continuously decreasing. Unstable dynamics are usually caused by the action of positive feedback (⁺feedback) processes. For example, the first-order ⁺feedback loop in Figure 1.2 informs us that a change in prey numbers is positively, or directly, related to prey numbers; i.e. the more prey there are the more there will be in the future, and vice versa. As long as this condition holds, the prey population will continue to change in the same direction, like the human or blue whale populations. Hence, ⁺feedback is the general process underlying the inflation spiral, the arms race, the population explosion, the extinction of species, and organic evolution.

1.2.2 Exogenous dynamics

Although the feedback structure of a dynamic system determines its properties of stability and instability, exogenous factors like temperature, rainfall, soil type, topography, etc. set the stage on which these dynamic interactions occur. For example, gradual increases in temperature due to global warming can change the interaction structure of ecological systems, say by making plants more susceptible to disease, and this can lead to instability due to outbreaks of disease. In contrast, normal variations in climate (weather) merely disturb the system temporarily from its stable state. Thus, we recognize two major kinds of exogenous effects: (1) those that cause changes in the stability properties of the ecosystem, which we call exogenous *forcing* processes; and (2) those that merely displace variables from their steady states and do not affect their stability, which we call exogenous *disturbances*. The latter are often considered to be *random* processes.

Because the feedback structure determines the stability properties of dynamic systems and, through this, the patterns and regularities that we observe in nature, it is important to understand how feedback loops are created in ecological systems, and how to detect and manipulate these feedback loops to produce stable, self-sustaining ecosystems. For this reason, the general principles of population dynamics developed in Part One of this book are built around these fundamental ideas. Then, in Part Two, methods are developed for detecting the feedback structure of real population systems, and for modelling these feedback processes. Finally, in Part Three, the methods are applied to a number of specific examples. In taking this approach we must accept the fact that ecological systems are extremely complex and, because of this, will never be completely understood nor precisely modelled. Each ecosystem is, to some extent, unique. Yet the fact that each person is unique does not prevent the physician from practising medicine, and the same facts should not deter the ecologist from analysing and prescribing treatments for natural ecosystems. Like the family doctor, the ecologist must use all the available scientific information and technology to make intelligent appraisals of probable cause (diagnosis) and possible remedial treatment (prescription). It is this basic philosophy that underlies the approach taken in this book.

1.3 SUMMARY

In this chapter we

1. Discussed some of the patterns and rhythms observed in natural populations, including unstable growth and decline, saw-toothed and cyclical oscillations, and shifts from one pattern to another.
2. Defined a theory as a systematic statement of the principles, processes and relationships underlying a natural phenomenon.
3. Showed why theory is necessary for understanding the causes of population changes and for intelligently managing populations of living organisms.

4. Demonstrated the effects of endogenous (feedback) processes, including stability (induced by ⁻feedback), cycles (induced by delayed ⁻feedback), and unstable dynamics (induced by ⁺feedback).
5. Demonstrated the potential effects of exogenous (non-feedback) processes, including forcing factors that can change or destroy the feedback structure and random processes that disturb systems from their steady states.
6. Discussed the general philosophy and approach taken in this book.

1.4 EXERCISE

Deer feed on certain woody shrubs and in so doing can severely reduce their abundance. Shrubs and trees compete with each other for space, sunlight, nutrients and water. Therefore, the more trees the less shrubs and vice versa. Shrubs, of course, are good for deer, but so are trees, for they provide protection. Draw the feedback structure of this system. How many feedback processes are there and are they negative or positive? Describe these feedback processes and explain how this system can be stabilized.

Population:
the central concept

2

To search for the best concept is no idle conceit, because the experiments that a scientist may devise and therefore the facts he may discover, as well as the explanations that he offers for them, depend on how he conceives nature. (H. G. Andrewartha and L. C. Birch, 1984)

Given that theory is necessary to understand and intelligently manage nature, then the next question to ask is "What should this theory be based upon?" Should ecological theory evolve around the idea of a population, or would it be better to base it on individual organisms, communities, or ecosystems? The fact of the matter is that ecological landscapes are really made up of individuals; individual trees, shrubs, deer, birds, insects and so on; multitudes of different individuals all going about their daily business and, in so doing, affecting each other in a multitude of different ways. Surely then a realistic theory of ecology should revolve around the individual organism and its innate genetic characteristics? The problem is that, although an individual-based theory may seem logical, it is not practical. First there is a computational problem, for if every organism in a large ecosystem were to be recognized as a separate entity, with its own particular character and behaviour, then keeping track of all the information, motion and interaction would be an impossible task, even for modern supercomputers. The second problem is one of measurement, for in order to forecast the future states of ecological variables we must first measure their present states. If an individual-based approach were to be employed, the state and location of every organism in the ecosystem would have to be measured before one could make a prediction! How can a pest manager measure the exact location, reproductive potential, and genetic make-up, of every insect in a field or forest? Thus, although an individual-based approach to ecology[4] may make sense, it is not practical.

One way to make a theory more practical is to work at a higher level of abstraction. The next level of ecological abstraction, above the individual, is the population. In this case we concern ourselves, not with the properties of individual organisms, but with the average properties of groups of individuals. Although variability from this average may be described statistically, say by the variance around the mean, the identity and uniqueness of the individual organism is lost in the abstraction. Thus, information about individual organisms is sacrificed in order to develop a manageable and practical theory. It should always be remembered, however, that a theory based on the characteristics of populations, although being more practical, is always less realistic or more abstract, than a theory based on individual organisms. On the other hand, a population-based theory is more realistic than a theory based on the idea of a community or ecosystem because it is a lower level of abstraction and, therefore, closer to reality.

2.1 THE POPULATION CONCEPT

Accepting the proposition that the population is the most practical unit for studying ecological dynamics, it is essential to clearly define what is meant by this term. Probably the most commonly used definition is *a group of individuals of the same species that live together in the same place*. This definition recognizes that populations are made up of individual organisms, but does not require us to know which individuals give birth or die, or where they are located in space. Instead the population is characterized by average birth and death rates, and variability in these averages is treated as a statistical property of the population.

The concept of population has certain similarities to that of *the species*. The two ideas are obviously interrelated and interdependent because a population is defined as a group of organisms of the same *species*, while a species is defined as a *population* of reproducing organisms that is reproductively isolated from other similar populations. The population is to ecology what the species is to systematics, the basic unit upon which the science is built. In addition, because the population is a central concept in systematics, evolution and ecology, it serves as a bridge to integrate and unify these areas of biology.

2.1.1 The spatial context

The definition of a population refers to individuals "living in the same place". Here place implies some kind of spatial resolution, as in "the population of New York" or "the population of aphids in a wheat field". But is "New York" or a "wheat field" an appropriate spatial scale in which to view human or aphid population dynamics[5]?

Of course, spatial dimension can be defined rather arbitrarily, say by the preferences (or prejudices) of the individual observer. An informal approach such as this is inherent in statements like "the population of aphids in a wheat field". However, in order to build a general and robust theory of population dynamics, one that applies to all species, it would be well to start with a formal and precise definition of the object of study. A theory based on arbitrary or confused concepts is likely to be an arbitrary and confusing theory! Perhaps this is one of the reasons why there is so much controversy and confusion in ecology?

Spatial dimension is one of the most difficult problems to solve when trying to define a particular population. For example, is a "wheat field" an appropriate spatial scale in which to study an aphid population, for aphids can move long distances, and their numbers can be affected by conditions in perennial habitats, far from wheat fields? As aphids invariably migrate into the farmer's fields from these perennial habitats, the correct spatial scale to study their dynamics must include these habitats, and may, therefore, be of the order of "dozens of square kilometres" rather than "a field". Using a smaller area may give rise to misleading or even dangerous inferences. For example, the conclusion that an aphid population is extinct because no aphids are present in a field is obviously a dangerous assumption if you are a farmer, for they may arrive in numbers tomorrow!

The problem, of course, is to determine the correct spatial scale, and to locate the correct boundaries, for the population under study. This problem can be tackled by considering two very different kinds of organisms – bacteria and elk. Bacteria are minute creatures with limited capability for movement, while elk are large animals that range over huge areas. It may make sense to study a population of bacteria on a square metre of ground, because an area this size will contain many thousands of individuals, and bacteria cannot move very far in an active sense. But it makes no sense to study a population of elk on the same square metre! Thus, the correct spatial scale for viewing a population depends on the nature of the organism. In general, large organisms, or organisms that can move long distances, will need to be studied over larger areas than small or sedentary organisms.

Probably the best rule of thumb is to choose an area that minimizes or balances the rate of *emigration* (or movement out of the area) and *immigration* (or movement into the area). Imagine the elk population, for example: if the area is too small, elk will be absent from it most of the time but occasionally the area may be crowded with animals. Almost all of the changes in the "population" will be due to movement into and out of the area. As the size of the area is increased, however, the rate of movement in and out will decline, and changes in numbers will be associated more and more with births and deaths rather than movements. When the elk population is correctly bounded, immigrations and emigrations should approach a minimum and be roughly balanced so that the variation in population numbers is almost entirely determined by births and deaths.

It is sometimes quite obvious where the actual boundaries of a population should be placed on the landscape. In the case of elk, for example, a specific region within which the population resides can often be identified; for example, an area surrounded by mountains, rivers, etc. With bacteria, on the other hand, it may be difficult or impossible to identify isolated populations and, in this case, the placement of boundaries may become rather arbitrary. However, the physical size of the area containing the population should still be large enough so that changes in numbers are associated more with births and deaths than with movements. The size of the area of study should, therefore, roughly correspond to the mobility of the organism, with more mobile organisms being studied over larger areas than sedentary ones.

It is now possible to provide a more precise definition, i.e. a *population is a group of individuals of the same species that live together in an area of sufficient size that they can carry out their normal functions, including migration, and where emigration and immigration rates are roughly balanced*. On occasion it may be necessary to talk about groups of organisms occupying smaller areas than the *true* population. When this occurs they will be identified as *sub*-populations or *local* populations. Local populations are characterized by unbalanced immigration and emigration rates and sometimes by local extinction. On the other hand, it may also be necessary to consider larger areas containing two or more distinct populations that occasionally share migrants. Groups of populations such as these will be identified as *meta*-populations.

2.2 POPULATION VARIABLES

Having defined a population, it is now necessary to determine which *variables* will be used to describe populations of living organisms, and how these variables can be estimated in the field. In the case of elk, for example, it may be possible to count every individual using aircraft or ground surveys at winter feeding sites. In this way one can obtain an accurate estimate of the total, or *absolute*, population present at a given time in a given area (the numbers of humans shown in Figure 1.1 are also absolute estimates). The absolute number can also be divided by the area of land within which the population resides to obtain an estimate of population *density* per unit area (e.g. gypsy egg masses are reported as numbers per hectare in Figure 1.1). Population density is a particularly useful variable because it relates population numbers to a standard area, say a hectare, and this enables us to compare populations of the same species inhabiting different areas, or populations of different species occupying the same area. Density also relates numbers to a constant measure of resource availability, because food and space are often interrelated and interdependent. Sometimes organisms may live in specialized habitats and, in such cases, densities can be measured in terms of habitat units, such as insects per square metre of foliage or soil (e.g. larch budmoths in Figure 1.1 are related to kilograms of larch foliage).

In contrast to elk, it is virtually impossible to count all the bacteria in a square metre of soil, or all the aphids over several square kilometres. In these cases, it is usually necessary to develop a *representative* sampling procedure; i.e. a sampling scheme that attempts to represent all the variability in the population, in much the same way as an opinion poll attempts to represent the variability in people's feelings about a particular issue[6]. The simplest way to obtain an unbiased estimate of population density in a given area is to sample at random points in space – a *random sample*. However, a more efficient procedure is to take a *stratified* random sample: in this case the variation in population density over the area of interest is first studied, and then a sampling protocol is designed that takes this variability into account; for example, the total area can be divided into regions, or strata, representing different densities, say high, medium and low, and then samples can be taken randomly within each stratum. Stratified random sampling enables one to obtain unbiased estimates of the mean density of the population per unit area of ground or habitat with minimal cost and effort. This estimate can also be converted into an absolute population count if the total area within which the population resides is known; i.e. the total number of hectares or square metres of soil. However, density estimates obtained by sampling are often preferable to absolute estimates, particularly if difficulties are encountered in defining the boundaries of the population.

On occasion it may be possible to census populations at special places. Sandhill cranes, for example, gather on the Platte River in Nebraska each spring, and elk can be monitored at winter feeding stations. On other occasions population counts can be obtained by monitoring hunters and fishermen (e.g. captures of blue whales and whitefish in Figure 1.1) or by trapping the organism (e.g. flying aphids can be captured by suction traps and moths by light traps). However, population counts obtained in this way are not necessarily related to area or absolute numbers, in which case they are called *indexes* of abundance.

Finally, population numbers are sometimes measured by what is called the *mark–recapture* method. Here individuals are live-trapped, marked in some way, released, and then trapped again. The proportion of marked to unmarked individuals can then be used to obtain an index of abundance. If certain conditions are met, it may be possible to transform this index into an estimate of absolute numbers or density.

2.3 POPULATION CHANGE

This book is concerned with changes in populations of living organisms over time. For this reason it is important to consider the correct time scale for observing population change. In other words, how long and how frequently should the population be observed (sampled) before analysis can lead to sensible conclusions about the causes of the observed population fluctuations? On the first question, one rule of thumb is that the time scale should be at least twice as long as the period of fluctuation. Thus, if the population exhibits 10-year cycles, as many do (see the larch budmoth in Figure 1.1), then at least 20 years of observations would be required. On the other hand, if populations oscillate at higher frequencies, like the sycamore aphid (Figure 1.1), then they may not need to be studied for so long to obtain meaningful data. Unfortunately, observations of population change over long periods of time are not always available to natural resource managers, and when they are they are often of poor quality. The fact is, managers must often make do with what is available, even if the data are inaccurate or sparse. The stance taken in this book is that any data are better than none. It is also important to realize that data sets get longer (and usually better) with time, but for this to happen someone must start collecting the data and others must keep it going.

On the second problem, the frequency of observations, a good rule of thumb is to observe the system at its natural frequency. On the planet Earth, the underlying natural rhythm is set by the revolution of the Earth around the Sun – the annual cycle. If one observes the dynamics more frequently than this, then most of the changes in population numbers may be caused by seasonal effects (exogenous forcing) rather than feedback, and it may be difficult or impossible to determine the important stabilizing or destabilizing mechanisms. In some cases it may be necessary to observe population changes more frequently. This is particularly true in agricultural systems, where the seasonal build up of pests is a critical problem. Because this book is more about natural or semi-natural situations, it is usually assumed (unless otherwise noted) that the populations being studied are measured at roughly the same time each year, and that this string of annual observations forms the *time series* describing the fluctuations of the populations over time – its dynamics.

2.3.1 The equations of change

Given a time series describing fluctuations in population density over time, then the change in density over a year can be represented by

$$\Delta N = N_t - N_{t-1},\qquad(2.1)$$

where N_t is the density of the population in year t and N_{t-1} is its density at the same time in the previous year. Of course, the change in a population of organisms over a year can also be measured if the numbers of births and deaths are known (assuming that immigrations and emigrations are balanced, as they should be if the spatial dimension of the population is properly defined). Hence, population change is also given by

$$\Delta N = \text{Births} - \text{Deaths}. \tag{2.2}$$

For reasons that will become clear later, it is preferable to express population birth and death rates as relative or *per capita* rates, so that population change over the year is written

$$\Delta N = BN_{t-1} - DN_{t-1} = N_{t-1}(B - D), \tag{2.3}$$

where B and D represent the per capita birth and death rates, respectively[7].

If the right-hand side of equation (2.1) is inserted for ΔN in equation (2.3), then

$$N_t - N_{t-1} = N_{t-1}(B - D), \tag{2.4}$$

and, rearranging so that only N_t appears on the left-hand side,

$$N_t = N_{t-1} + N_{t-1}(B - D), \tag{2.5}$$

or

$$N_t = N_{t-1}(1 + B - D). \tag{2.6}$$

This is called a *step-ahead forecasting equation* because population numbers 1 year into the future can be forecast from their current numbers if the per capita birth and death rates are known.

The quantity $1 + B - D$ in equation (2.6) measures the per capita rate of change over a finite period of time, say a year, and is sometimes called the *finite per capita rate of change* of the population. This quantity will be represented by the variable G (the Greek symbol λ is also used in the literature)

$$G = 1 + B - D, \tag{2.7}$$

so that equation (2.6) becomes

$$N_t = N_{t-1}G. \tag{2.8}$$

As will become clear later, it is often advantageous to express the growth rate of a population in natural logarithms; i.e.

$$\ln N_t = \ln N_{t-1} + \ln G, \tag{2.9}$$

where ln represents the natural logarithm of the quantity that follows it. Defining the *logarithmic per capita rate of change* as

$$R = \ln G = \ln(1 + B - D), \tag{2.10}$$

equation (2.9) becomes

$$\ln N_t = \ln N_{t-1} + R. \tag{2.11}$$

Notice that if estimates of N_{t-1} and N_t are available, then the per capita rate of change, R, can be estimated from the relationship

$$R = \ln N_t - \ln N_{t-1} = \ln\left(\frac{N_t}{N_{t-1}}\right) \tag{2.12}$$

R is often called the *instantaneous* or *intrinsic* rate of increase of the population, and is often given the lower case symbol r. In this book, however, the symbol r is reserved for the correlation coefficient and so the upper case is used for the logarithmic rate of change.

2.4 SUMMARY

1. The concept of population is an abstraction that is necessary for practical reasons.
2. The correct spatial scale for viewing a population depends on the size and mobility of the species. The best rule of thumb is that the spatial scale be such that emigration and immigration rates are roughly balanced.
3. A population is defined as a group of individuals of the same species living together in an area of sufficient size that they can carry out their normal functions, including migration, and where emigration and immigration rates are roughly balanced.
4. The state of a population at a given instant in time is described by the population variables absolute numbers, density, or indexes of abundance.
5. Variables can be measured by total counts, random or stratified random samples, harvest records, traps or mark–recapture techniques.
6. The longer the series of observations on population changes (time series) the better, but we must often make do with what we have.
7. In order to avoid the complications associated with exogenous seasonal effects, population dynamics is usually studied on estimates of population size taken at the same time each year.
8. Rates of change are described in terms of per capita birth and death rates and the finite and instantaneous per capita rates of change (G and R).

2.5 EXERCISES

1. 758 elk spend 3 months each winter in a 5 hectare feed lot. These elk are known to range over a watershed of 1230 hectares. What is the absolute size of the population and what is its density?

2. A scientist sampled the white grubs in a 3.4 hectare pasture by taking 200 randomly located 1 decimetre (= 100 square centimetres) soil cores. He counted 7564 white grubs in these samples. Calculate the density of white grubs per square metre of soil, per hectare, and the absolute population in the pasture.

3. The elk population discussed above produced 58 calves in 1990 but only 22 were alive by 31 December 1990. In addition, 37 yearling and older elk died during 1990. What was the per capita birth and death rate and the logarithmic net per capita rate of change (R) over the year? Is this elk population increasing?

3

The first principle: exponential growth

I think I may fairly make two postula. First, that food is necessary to the existence of man. Second, that the passion between the sexes is necessary and will remain nearly in its present state. Assuming then, my postula as granted, I say, that the power of population is indefinitely greater than the power in the earth to produce subsistence for man. (T. R. Malthus, 1798)

Malthus' *Essay on the Principle of Population* was one of the earliest explorations into the theory of population dynamics. What Malthus realized is that populations of humans, and in fact any self-replicating entity, can grow exponentially or geometrically, and that this kind of growth can overwhelm finite or arithmetically increasing resources. Exponential growth is most easily visualized by considering a single amoeba which, we assume, reproduces by division once every day. After 2 days there will be two amoebas, after 3 days four, then eight, then 16, then 32, then 64, then 128 and so on. In 10 days there will be 1024 amoebas, in 20 days a million, and in a month a trillion. As Malthus realized, exponential population growth is indeed a powerful force.

Exponential growth will be known henceforth as the *first principle of population dynamics*. By using the term "principle" we imply that exponential growth is a fundamental, if somewhat obvious, property of all population systems.

3.1 MATHEMATICAL INTERPRETATION

The first principle can be derived mathematically from the step-ahead forecasting equation (2.8)

$$N_t = N_{t-1}G. \tag{3.1}$$

where $G = 1 + B - D$ is the finite per capita rate of change, with B and D the per capita birth and death rates, respectively. This equation is called a finite *difference equation* because it computes the growth of the population over a finite period of time, like a year or a generation.

The growth of a population can be computed from equation (3.1) by starting with a population of N_0 individuals at time zero. After one time period, say a year, there will be

$$N_1 = N_0 G, \tag{3.2}$$

individuals, after 2 years

$$N_2 = N_1 G, \tag{3.3}$$

and so on.

Substituting the right-hand side of equation (3.2) for N_1 in equation (3.3) we have

$$N_2 = (N_0 G)G = N_0 G^2. \tag{3.4}$$

In fact it is possible to solve the equation for any length of time t

$$N_t = N_0 G^t. \tag{3.5}$$

to arrive at a finite difference equation describing the first principle of population dynamics.

The first principle can also be derived as a *differential* or continuous-time equation. For example, the *instantaneous* rate of growth of a population, dN/dt, is obtained from the instantaneous per capita rate of change, R, multiplied by the density of the population, N, so that

$$\frac{dN}{dt} = RN. \tag{3.6}$$

Note that the instantaneous rate of change is measured when the time step is very small (i.e. when $dt \to 0$). Rearranging equation (3.6) so that only N appears on the left-hand side

$$\frac{1}{N} dN = R dt, \tag{3.7}$$

and integrating over time, yields

$$N_t = N_0 e^{Rt}, \tag{3.8}$$

where e is the base of the natural logarithm (e $= 2.71828\ldots$). Converting to natural logarithms gives the linear relationship

$$\ln N_t = \ln N_0 + Rt. \tag{3.9}$$

Transforming equation (3.5) to natural logarithms,

$$\ln N_t = \ln N_0 + (\ln G)t, \tag{3.10}$$

we see that equations (3.5) and (3.8) are identical with

$$R = \ln G. \tag{3.11}$$

Equations (3.5), (3.8), (3.9) and (3.10) are all precise mathematical statements of the first principle of population dynamics. This principle is truly general in that it applies to all populations and, in fact, to all self-replicating entities, including your compound interest savings account. It is sometimes called the "law of population growth" or the "Malthusian law". These equations will be used, in one form or another, whenever we want to calculate the *trajectory* that a population takes over time, or to predict the future growth of a population when the per capita rate of change and current population size are known.

3.2 POPULATION DYNAMICS UNDER THE FIRST PRINCIPLE

Let us examine the dynamic consequences of the first principle by calculating the growth of a population obeying equation (3.8). First write equation (3.8) in step-ahead forecasting form by setting the time step to unity ($dt = 1$) so that

$$N_t = N_{t-1}e^R. \tag{3.12}$$

Then set the initial population density at $N_0 = 10$ and assume an average per capita rate of change of $R = 0.5$ per year. After 1 year the population will be

$$N_t = 10e^{0.5}.$$

The value of $e^{0.5} = 1.6487$ can be found in a table of exponents or from a pocket calculator, and so

$$N_1 = 10 \times 1.6487 = 16.5$$
$$N_2 = 16.5 \times 1.6487 = 27.2,$$

and so on. This is what is called a *numerical* solution of the growth equation, or a simulation of the growth process. Numerical solutions for three different values for R are shown in Figure 3.1. As expected, populations with $R > 0$ (or the birth rate greater than the death rate) grow at an increasing rate (exponentially), those with $R = 0$ (or the birth rate equal to the death rate) remain unchanged, while those with $R < 0$ (or the birth rate less than the death rate) decay exponentially towards zero (Figure 3.1, left). If the logarithm of population density is plotted against time, a straight line or linear relationship is obtained with the slope equal to the per capita rate of change (Figure 3.1, right). [Note that logarithmic transformation of the exponential growth curve is a straight line as shown by equation (3.10).]

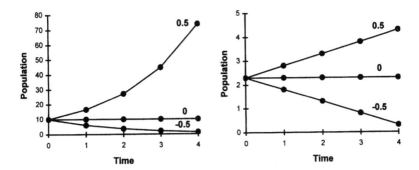

Figure 3.1 Simulation of population growth according to the first principle; equation (3.12) with $N_0 = 10$ and $R = 0.5$, 0, and –0.5, plotted on the arithmetic (left) and logarithmic (right) scales.

Populations growing or declining exponentially are said to be *unstable* because they tend to move away from their original, or initial, conditions. They are also said to be exhibiting *transient* or *non-equilibrium* dynamics – they are in transition, or in the process of going somewhere else, perhaps to extinction. The reason for this instability is that the first principle defines a first-order †feedback process because larger populations give rise to higher growth rates; i.e. the rate of population growth is determined by the product RN [see equation (3.6)]. Notice, however, that the population can be in equilibrium if the exact condition $R = 0$ is met, or the birth rate exactly equals the death rate, but this condition is unlikely to occur for very long in nature where variation is the norm.

3.2.1 Sensitive dependence

One of the characteristics of systems governed by †feedback growth processes is that their current states depend critically on their initial, or starting, states. This is called *sensitivity to*, or *sensitive dependence on*, initial conditions. Let us illustrate this with the differential equation (3.8). Suppose there are two isolated populations, both with the same per capita rate of change ($R = 0.5$) but one containing 10 individuals and the other 11. After five time steps the first population will have 316 individuals and the second one 401, a difference of 85. However, by the 10th time step the difference will be 61,051 and by the 20th this difference will have grown to over 15 billion. In fact it is easy to show that the initial difference between the two populations grows exponentially with time[8]. The phenomenon of sensitive dependence is important, for it means that small errors made when estimating population size can be amplified with time so that long-range predictions become almost impossible, at least for exponentially expanding populations.

3.2.2 Stochastic growth

Populations in nature rarely exist in constant environments. In order to simulate more realistic growth trajectories, some random variation, or *noise*, can be included in the calculations. This is done by adding a random variable, sZ, to the value of R, so that equation (3.12) becomes a *stochastic difference equation*

$$N_t = N_{t-1} e^{R+sZ}, \qquad (3.13)$$

where s, the *standard deviation*, defines the amount of random variability in the value of R, and Z is a *random normal deviate* (see Appendix). A random normal deviate is a value selected at random from a *normal distribution* with mean $\bar{Z} = 0$ and standard deviation $s = 1$. Because the mean of the distribution is zero, most of the numbers chosen will be close to zero and, therefore, the value of $R + sZ$ will usually be close to R. The probability of selecting large negative or positive numbers is directly proportional to the standard deviation.

Let us now examine transient dynamics in a noisy world: starting with 10 organisms, an average per capita rate of change 0.5, and standard deviation 0.1, we read the first value of $Z = 0.614$ from the table of random normal deviates (Appendix; pick a number with your eyes closed). Inserting these values in equation (3.13) gives us

$$N_1 = 10 \times e^{0.5+0.1 \times 0.641} = 17.5,$$

then, taking the next random deviate, $Z = -0.66$,

$$N_2 = 17.5 \times e^{0.5+0.1 \times (-0.66)} = 27,$$

and so on. Figure 3.2 illustrates the transient dynamics of a population inhabiting a noisy environment. Notice that the population still grows exponentially and, when plotted as logarithms, retains its basic linear form.

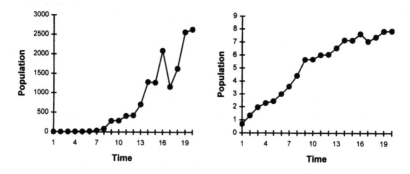

Figure 3.2 Population growth according to the stochastic equation (3.13) with parameters the same as in Figure 3.1 and $s = 0.1$, plotted on arithmetic (left) and logarithmic (right) scales.

3.3 SUMMARY

In this chapter we presented

1. The first principle of population dynamics, which states that populations grow and decline geometrically or exponentially when their rates of change are constant.
2. On the logarithmic scale, the growth of populations obeying the first principle is linear with the slope equal to the average per capita rate of change.
3. The basic forecasting equations were derived in terms of finite per capita rate of change, G (i.e. $N_t = N_{t-1}G$) and the instantaneous per capita rate of change, R (i.e. $N_t = N_{t-1} e^R$) with $R = \ln G$.
4. The first principle defines a first-order +feedback loop because the rate of change of the population is directly related to its own size.
5. This +feedback loop gives rise to unstable dynamics; accelerating growth when $G > 1$ or $R > 0$, decay to extinction when $G < 1$ or $R < 0$, and equilibrium under the unlikely condition that $G = 1$ or $R = 0$.
6. Exponential growth processes are sensitive to their initial conditions so that accurate long-term forecasts are virtually impossible.
7. Numerical solutions show that these transient patterns are manifested in both constant and noisy environments.

3.4 EXERCISES

1. If you have \$10,000 in the bank at 10% annual interest rate, how much will you have in 10 years if the interest is compounded at the end of each year? What is the relationship between the interest rate and the population parameters G, R, B and D?
2. Simulate the dynamics* of a population for 10 years using equation (3.13) when the initial density is 10 individuals, the per capita rate of change is $R = 0.8$ and the standard deviation is zero. Plot the trajectory on both arithmetic and logarithmic scales.
3. Simulate the dynamics* of a population governed by equation (3.13) under the same conditions as before but with the standard deviation 0.2. Plot the trajectory on arithmetic and logarithmic scales.

*These problems can be done by hand or on a typical computer spreadsheet.

<table>
<tr><td>

4

</td><td>

The second principle:
co-operation

</td></tr>
</table>

It is sufficient to state here that aggregations formed without sexual stimuli and at the lowest level of group integration may have survival value for their members and, under certain conditions, are essential for the survival of the race. (W. C. Allee, 1931)

The first principle of population dynamics applies directly to all reproducing entities with *constant* rates of change. The problem, of course, is that birth and death rates do not normally remain constant in natural populations. The next three principles of population dynamics define the main feedback processes that affect the birth and death rates of individual organisms.

4.1 THE *R*-FUNCTION

Before discussing the second principle of population dynamics, it is necessary to introduce the concept of a feedback function and, more specifically, the reproduction function or *R-function*. As we saw in Chapter 1, the properties of stability and instability of dynamic systems depend on their endogenous feedback structures, with ⁻feedback acting to stabilize and ⁺feedback to destabilize dynamics. Remember that feedback occurs whenever a change in the state of a dynamic variable is affected by past values of that variable. In other words, if the variable of interest is population density, N, then a *density-induced* feedback process is defined by

$$\Delta N = F(N_{t-d}), \tag{4.1}$$

where ΔN represents a change in population density over a finite unit of time, usually a generation or year, and $F(N_{t-d})$ is an arbitrary function of population density d time units (generations or years) in the past. The function F, therefore, is a *feedback function* that defines how a population changes in response to previous densities. Now the first principle of population dynamics meets this definition because, given equation (3.1) with time step of unity, then

$$\Delta N = N_t - N_{t-1} = N_{t-1}G - N_{t-1} = N_{t-1}(G - 1) = F(N_{t-1}). \qquad (4.2)$$

This equation describes a $^+$feedback process because changes in N over a unit of time are positively related to population density at the beginning of the time period, N_{t-1}. If population change is calculated from the natural logarithm of population density, however,

$$\Delta \ln N = \ln N_t - \ln N_{t-1} = \ln N_{t-1} + \ln G - \ln N_{t-1} = \ln G = R, \qquad (4.3)$$

and we see that logarithmic population growth is independent of population density. This is an important observation because the first principle is universal, in that it applies to all populations all of the time, so that there is no need to diagnose its presence or absence. On the other hand, other feedback processes are neither universal nor constant, so it is important to be able to determine if they are operating. Because the $^+$feedback due to the first principle is removed by the logarithmic transformation, any relationship observed between the logarithmic rate of change, R, and past population densities, N_{t-d}, must be due to these other feedback processes. In other words, we can detect and analyse non-trivial feedback by examining the relationship

$$R = \ln N_t - \ln N_{t-1} = f(N_{t-d}). \qquad (4.4)$$

where R, no longer a constant, is now called the *realized* per capita rate of change of the population. Notice that equation (4.4) is a feedback function, as defined by equation (4.1), because when R is a function of N so too is the rate of population change ΔN; i.e. $R = \Delta \ln N$ as seen in equation (4.3). Because the function f describes how the net reproductive rate of the average individual changes with population density, it has been called the *reproduction* function or, for short, the *R-function*.

The *R*-function is an extremely important concept in population dynamics theory. It defines how the well-being or *fitness* of an average individual changes with the density of the population within which it lives; i.e. a large positive value of R means that most members of the population live a long time and produce many offspring, while a large negative R means that most individuals die before giving birth. To coin an economic analogy, R represents the "standard of living" of the average individual, and the *R*-function describes how the "standard of living" changes with population density (we will see many more analogies between ecology and economics in this book). The *R*-function also describes the feedback structure that *regulates* (or controls) the dynamics of a population of living organisms and, for this reason, is sometimes called the *regulation function*. When R is a direct (positive) function of population density, the R-function describes a +feedback process because an increase in population density results in higher per capita rates of change, and this causes the population to increase still further. On the other hand, when R is an inverse (negative) function of population density, the R-function describes a ⁻feedback process because an increase in density causes the rate of change to decline and this results in a future decrease in population density. These general rules can be written as follows (see Figure 4.1):

+feedback when $\dfrac{\partial R}{\partial N} > 0$, or the slope of the R-function is positive   (4.5a)

⁻feedback when $\dfrac{\partial R}{\partial N} < 0$, or the slope of the R-function is negative   (4.5b)

no feedback when $\dfrac{\partial R}{\partial N} = 0$, or the slope of the R-function is zero.   (4.5c)

Note that $\partial R/\partial N$ is the partial derivative of R with respect to N and that it represents a change in R with respect to a change in N, or the slope of the R-function.

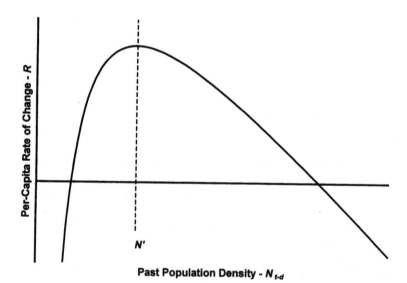

Figure 4.1 Hypothetical R-function showing the relationship between the per capita rate of change, R , and past population density, N_{t-d}. The curve is divided into two parts by the dotted line, N': To the left of the dotted line +feedback dominates because $\partial R/\partial N > 0$ (+ slope), to the right ⁻feedback dominates because $\partial R/\partial N < 0$ (– slope), and at the peak of the curve there is no feedback because $\partial R/\partial N = 0$ (0 slope).

4.2 CO-OPERATION

Having introduced the concept of the R-function, we can now return to the *second principle of population dynamics*. The second principle recognizes that individual organisms can aid each other in the struggle to survive and reproduce and that this can affect the shape of the R-function. For example, the second principle is expressed whenever organisms form into herds, packs, schools, flocks, and swarms that help them to obtain resources or avoid enemies. Because aggregations are more likely to form when populations become dense, the benefits received from the second principle often increase with population density, and this can give rise to +feedback through the R-function (i.e. $\partial R/\partial N > 0$ or the R-function has a positive slope). Perhaps the most obvious example is the increased probability of encountering a mate, and hence of reproducing, in large populations. Not so obvious is the fact that individuals are less likely to be killed by predators when they are in a crowd. Think of it this way: if a single wildebeest wanders into the territory of a hungry lion pride its chances of being eaten will be quite high, but if it is in a herd of thousands its chance of being eaten will be very small. For the individual there is indeed "safety in numbers".

The second principle is analogous to the economic *law of increasing returns*, which recognizes that organizations can become more productive as they get larger because coalitions between individuals, departments, and such, make the organization more efficient.

When individuals benefit from the presence of others, they are said to *co-operate* with each other. Co-operation between members of the same species is called *intra*specific co-operation and that between different species *inter*specific co-operation. There are two basic ways in which co-operation can arise: in the first, the co-operating organisms possess adaptive traits that have evolved over time to ensure the cohesion of the group during co-operative hunting or defence activities. Many of these co-operative adaptations involve some kind of *social behaviour* that facilitates group interactions. We will call this kind of co-operation *adapted* or *programmed*. Adapted co-operation has evolved in many species, including the social insects (bees, ants, and wasps), fish and birds (schooling and flocking), mammals (herds, packs and prides), and also in our own species where group hunting has evolved into a culture of science and technology, the co-operative acquisition and application of knowledge. In addition to adapted co-operative interactions between members of the same species, co-operation can also evolve between different species of animals and plants. This is called adapted *inter*specific co-operation, or mutualism and symbiosis.

The second kind of co-operation results from accidental aggregations that coincidentally provide individuals with access to resources or avoidance of enemies. For example, just happening to be in a crowd may provide one with a greater chance of finding a mate and a lower probability of being eaten by predators. Accidental or unintentional co-operative activities will be called *unadapted* or *incidental* co-operation. This kind of co-operation can also occur between members of the same species (*intra*specific) or between members of different species (*inter*specific).

4.2.1 Underpopulation (Allee) effects

Consider what is likely to happen to individuals in a sparse population of sexually reproducing organisms as the density of the population gradually increases. For a start, the probability of meeting a mate should increase and, because of this, the per capita birth rate (B) will tend to rise towards a maximum value, the reproductive potential of the species, as population density increases (see curve B in Figure 4.2, top). This kind of *intra*specific co-operation could be either adapted, as in the evolution of mating swarms and colonies, or incidental. If we assume for the moment, and without loss of generality, that the death rate (D) is lower than the maximum birth rate and does not change with density (line D in Figure 4.2, top), then there will be a point U where the birth and death curves cross. At this point the population must be in equilibrium because births and deaths are exactly equal. Notice, however, that the death rate exceeds the birth rate when the population density is less than U (i.e. $B < D$ for all $N < U$ in Figure 4.2, top). Thus, if the population is below U it will decline until it eventually becomes extinct. This property of sparse sexually reproducing populations is called an *underpopulation* or *Allee* effect[9]. On the other hand, if the population is greater than U, births will exceed deaths and the population will grow continuously (because $B > D$ for all $N > U$). Notice that the direction of population change is given by arrows that point away from the equilibrium (Figure 4.2), indicating that U is an *unstable* or *divergent* equilibrium, sometimes called a *repeller*.

Given the relationship between the per capita birth and death rates and population density shown in Figure 4.2, then the R-function can be determined from equation (2.10); i.e. $R = \ln(1 + B - D)$ (Figure 4.2, bottom). Notice that the R-function created by *intra*specific co-operation during mating rises with population density up to a maximum value A. This is called the *maximum per capita rate of change*. Also note that the R-function generates +feedback when populations are sparse (because $\partial R/\partial N > 0$) and no feedback when populations get very large (because $\partial R/\partial N = 0$), and that the unstable equilibrium occurs at the point where the R-function crosses the line $R = 0$ (because $B = D$ when $R = 0$).

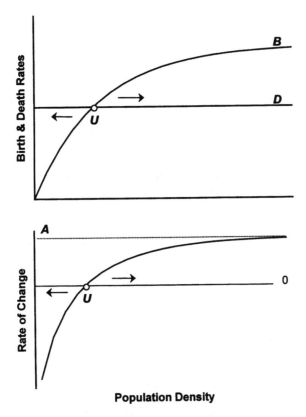

Figure 4.2 Top, relationship between population density and the per capita birth (B) and death (D) rates when B is assumed to rise with population density due to higher mating frequencies and D is assumed to be unaffected by population density. Bottom, the resulting relationship between $R = \ln(1 + B - D)$ and population density (N_{t-1}) – the R-function. Notice that U is an unstable equilibrium density occurring at the point $B = D$ or $R = 0$. Arrows signify the direction of population change in the vicinity of U.

4.2.2 Group defence

As we have noted previously, being in a crowd provides a certain degree of safety from attack by enemies. The reason for this is that most consumers have finite appetites, and stop feeding when they are *satiated*. Imagine a population of prey being fed upon by a single predator: when the prey population is sparse, the predator will always be hungry, and any increase in the numbers of prey will immediately result in an increased rate of attack by the predator. However, when prey are very abundant, the predator will always be satiated and an increase in prey density will have no effect on its rate of attack. Hence, the attack rate of a single predator will increase at first with prey density but will eventually level off at a constant rate (Figure 4.3, top). This relationship between the rate of attack of a single predator and the density of its prey is called the *behavioural* or *functional* response[10] of the predator. Behavioural responses such as that in Figure 4.3 (top) are called C-shaped or cyrtoid.

The probability of a prey being killed by predators with cyrtoid behavioural responses can be calculated by dividing the number killed (Figure 4.3, top) by the density of the prey population. Plotting this probability against prey density shows that the likelihood of prey death *declines* with prey density (Figure 4.3, middle); i.e. the "safety in numbers" effect. If births are assumed to remain constant, then the resulting R-function will have a positive slope at sparse prey densities and zero slope at very high prey densities (Figure 4.3, bottom). Like the Allee effect, cyrtoid behavioural responses generate +feedback on the prey population, and also give rise to unstable equilibria in the R-functions of their prey (Figure 4.3). For reasons that will become clear later, the unstable equilibrium created by the behavioural responses of predators is called an *escape threshold* and is identified by the letter E (Figure 4.3, bottom).

4.2.3 Group hunting

In some species, individuals co-operate with each other during the capture of prey or while gathering other resources. This is particularly true when prey are large and vigorous, or have effective defences. For example, wolf packs can kill large mammals like moose, while single wolves are restricted to eating small animals like beavers, rabbits and mice. Wolf packs, therefore, have access to a much larger food supply than single individuals. Bark beetles that attack and kill large pine trees are another example: certain species of bark beetles can kill healthy, vigorous trees when their numbers are high but are restricted to severely weakened or dead trees when their numbers are low (see Chapter 14 for more on bark beetles). Like wolves, large bark beetle populations have access to a much more extensive food supply. Organisms that co-operate with each other when attacking large and/or vigorous prey can also have R-functions with unstable equilibrium points because birth rates can increase (as in Figure 4.2), and/or death rates decrease (as in Figure 4.3), as population density rises.

4.2.4 Population thresholds

A major consequence of the second principle of population dynamics is that the R-function can have unstable equilibrium points that separate qualitatively different kinds of dynamic behaviour (e.g. growth from

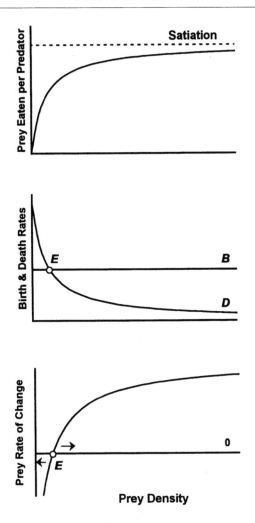

Figure 4.3 Top, the cyrtoid or C-shaped behavioural response of a predator showing how the number of prey killed per predator per unit of time changes in relation to prey density. Middle, the resultant effect on the per capita death rate of the prey (D), under the assumption of a constant prey birth rate (B). Bottom, the corresponding R-function for the prey with E an unstable equilibrium or escape threshold (in this case it is also an extinction threshold).

extinction behaviours). Because of this, the unstable equilibrium is sometimes called a *population threshold* or *breakpoint*. There are two main kinds of population thresholds. The first type arises from co-operative processes that manifest themselves at very low population densities, the most obvious being the positive relationship between the birth rate and population density due to increased mating success (Figure 4.2). If an equilibrium point is created under these circumstances, it usually separates extinction from growth dynamics and, for this reason, is often called an *extinction* or *underpopulation threshold* (*U* in Figure 4.2).

The second kind of population threshold arises from the behavioural responses of enemies or by group attack on large and/or vigorous prey. These manifestations of the second principle can occur at relatively high population densities and may allow populations to *escape* the constraints placed upon them by their enemies or food supplies. For this reason, the unstable equilibrium is usually called an *escape threshold* (*E* in Figure 4.3). Of course, escape thresholds may also be extinction points but, as we will find in Chapter 7, they may also give rise to more complicated *R*-functions.

4.3 MATHEMATICAL INTERPRETATION

There are three conditions under which the second principle of population dynamics can be evoked – mating success, the behavioural response of enemies, and group hunting. However, our mathematical interpretation will be developed around the idea of escape from enemies because this is where most of the theoretical and empirical research has been carried out. It should be realized, however, that the same basic mathematical interpretation applies to the other factors.

Consider the case of a wildebeest herd being fed upon by a pride of lions. Suppose that there are P lions and that each eats W wildebeest during a year. This will result in WP wildebeests being killed each year, and the per capita death rate of wildebeest living in a herd of size N will be

$$D_N = \frac{WP}{N}.$$

(4.6)

Notice that the per capita death rate of wildebeests declines as the herd size increases because the death rate is inversely proportional to the population density (Figure 4.3, middle).

Now suppose that there are also other species of prey in the area. Under these conditions, the per capita death rate of wildebeest will also depend on the density of these alternative prey and the relative preference of lions for wildebeest over these other prey; i.e.

$$D_N = \frac{WP}{N + F},$$

(4.7)

where F measures the availability, and relative preference, of lions for alternative prey. Notice that the other prey species act as incidental *inter-specific* co-operators because their presence lowers the per capita death rate of wildebeests. Hence, F also measures the degree of *interspecific* co-operation.

The behavioural response[10] of individual lions to prey density, or the numbers of prey killed per lion per year, N_K, can be obtained by multi-plying equation (4.7) by N and dividing by P

$$N_K = \frac{N}{P} \cdot \frac{WP}{N+F} = \frac{WN}{N+F}. \tag{4.8}$$

Note that if N_K is plotted against N we obtain a cyrtoid response similar to that of Figure 4.3 (top).

Assuming that the maximum per capita rate of change of the prey population in the absence of predation is A_N, and that this quantity is reduced in proportion to the per capita death rate due to predation, D_N, then

$$R_N = A_N - D_N = A_N - \frac{WP}{N_{t-1} + F}, \tag{4.9}$$

where R_N is the *realized* rate of change of the average *prey* individual in a population of N prey, and in an environment containing P predators and F alternative prey. Note that the use of the time subscript identifies N_{t-1} as a *dynamic* variable while all other quantities are constants. The parameter A_N, the maximum per capita rate of change of the prey, can be thought of as the genetic potential of the prey species when living in a particular environment that is free of predators. The parameter W is no longer the simple feeding rate of an individual predator, but now measures the reduction of the prey's per capita rate of change caused by the addition of a single predator to the environment (i.e. it is the orig-inal W multiplied by a proportionality constant). This parameter now recognizes that predators can affect prey reproduction as well as survival, say by interfering with mating or courtship behaviour. In other words, it measures the total *impact* of the predator on the per capita rate of change of the prey.

It is important to remember that the R-function defined by equation (4.9) describes the action of the second principle on a population of *prey* subjected to a constant rate of predation[11]. This is reflected by the subscripts of the R-function; i.e. R_N identifies the R-function as belonging to a population of prey, N. This is necessary because R-functions for consumer or predator populations, with subscript P, will be encountered in later chapters.

The escape threshold for the prey population occurs when $R_N = 0$ and $N = E$, and so substituting these values in equation (4.9) provides us with a mathematical definition of the escape threshold

$$0 = A - \frac{WP}{E + F}$$

(4.10)

$$E = \frac{WP}{A} - F$$

Notice that the escape threshold for the prey is directly related to the impact of the predator on the prey W and to predator density P, and inversely related to the maximum per capita rate of change of the prey A_N and the degree of *inter*specific co-operation F. In addition, when alternative prey are very abundant so that $F > WP/A$, then $E < 0$ and the escape threshold vanishes. Under this condition the prey population cannot be suppressed to very low densities, or eradicated, by its predator.

4.4 POPULATION DYNAMICS UNDER THE SECOND PRINCIPLE

The dynamics of populations obeying the second principle can be simulated with equations (3.12) and (4.9). For example, set the parameters of equation (4.9) to $A_N = 1$, $W = 10$, $F = 10$, $P = 11$ and $N_0 = 99$, and then calculate the value of R_N

$$R_N = A_N - \frac{WP}{N_0 + F} = 1 - \frac{10 \times 11}{99 + 10} = -0.0092.$$

Now insert this value into the step-ahead forecasting equation (3.12) to calculate the density of the prey population after one time step

$$N_1 = N_0 e^{R_N} = 99 e^{-0.0092} = 98.1.$$

Notice that the population of prey has declined, indicating that the starting population density was below the escape threshold, E. Of course, the threshold can be computed with equation (4.10)

$$E = \frac{WP}{A_N} - F = \frac{110}{1} - 10 = 100.$$

showing that $N_0 = 99$ is indeed just below the escape threshold. Continuing to calculate the decline in the population, N_1 is inserted into equation (4.9) to find $R_N = -0.0176$ and then this value is used to calculate $N_2 = 96.4$ from equation (3.12), and so on (Figure 4.4). This figure

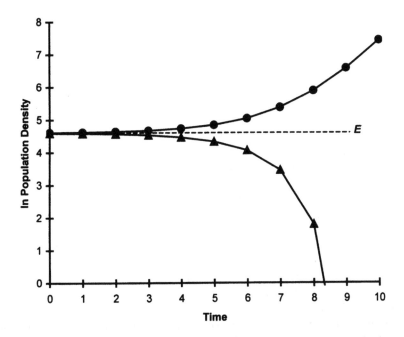

Figure 4.4 Population dynamics predicted by equations (3.12) and (4.9) when $A_N = 1$, $W = 10$, $P = 11$, $F = 10$ and $N_0 = 101$ (increasing trajectory) or $N_0 = 99$ (declining trajectory).

also shows a population trajectory starting from $N_0 = 101$, just above the escape threshold, E.

The main conclusion from the analysis of population dynamics under the second principle is that populations grow when they are above an unstable threshold and decline when they are below it. The thresholds, therefore, separate the dynamics into two qualitatively different behaviours (growth versus collapse). In addition, if the logarithm of population density is plotted against time, we see that the trajectories are curvilinear (Figure 4.4) rather than linear (remember that exponential population growth under the first principle is linear when plotted on the logarithmic scale). This is because population growth now involves two ⁺feedback processes, geometric growth plus co-operation, and this gives rise to *hyper-geometric* or *hyper-exponential* population growth and collapse. Hyper-geometric growth must eventually grade into exponential growth because there is always a finite maximum birth rate or a minimum death rate. For example, notice how the growth pattern in Figure 4.4 eventually becomes more or less linear. On the other hand, hyper-geometric collapses to extinction are particularly violent because the rate of decline accelerates with time. This observation should be of considerable significance to those interested in the conservation of endangered species, for the rush towards extinction can occur at an ever-increasing rate.

4.5 SUMMARY

In this chapter we

1. Defined the central concept of the *R*-function, which shows how the realized rate of change of the individual organism changes with respect to population density in the absence of +feedback due to the first principle.
2. Defined the second principle of population dynamics, which recognizes that co-operative aggregations may form when populations become large, and that these co-operative activities can give rise to higher per capita birth rates and/or lower death rates as populations become larger or denser.
3. Defined the concepts of *intra-* and *inter*specific co-operation and increasing returns, and the notions of adapted (programmed) and unadapted (incidental) co-operation.
4. Defined the underpopulation or Allee effect due to density-related mating success and showed how it affects the *R*-function and gives rise to +feedback on population growth.
5. Defined the cyrtoid behavioural (functional) response of a consumer and showed how it gives rise to +feedback in the prey (resource) *R*-function.
6. Showed how +feedback and escape thresholds can also be created in populations of organisms with group hunting and group defence behaviours.
7. Showed how the second principle creates unstable population (extinction and escape) thresholds and hyper-geometric population growth.

4.6 EXERCISES

1. Simulate the dynamics* of a population under the second principle using equations (4.9) and (3.12) with $A = 1, W = 10, P = 11, F = 10, N_0 = 101$, then with $N_0 = 99$. Calculate the escape threshold for $P = 11, 21, 41$.
2. Draw a graph of the *R*-function generated by the second principle using equation (4.9) with parameters given in the above exercise.

*This problem can be done by hand or on a typical computer spreadsheet.

The third principle: competition | 5

There is no exception to the rule that every organic being naturally increases at so high a rate, that if not destroyed the earth would soon be covered by the progeny of a single pair ... Battle within battle must ever be recurring with varying success; and yet in the long run the forces are so nicely balanced, that the face of nature remains uniform for longer periods of time. (C. Darwin, 1859)

The *third principle of population dynamics* recognizes that individuals living in dense populations may have difficulty acquiring the resources they need to survive and reproduce. This principle also has its economic analogy in the *law of diminishing returns*, which describes how firms and bureaucracies often become less efficient as they get larger. It is also similar to Malthus' idea of a *"struggle for existence"* which Darwin borrowed for his *"theory of evolution"*.

5.1 COMPETITION

The struggle between individual organisms to obtain the resources they need to survive and reproduce is usually called *competition*. As with co-operation, interactions between members of the same species are called *intra*specific while those between different species are called *inter*specific. Similarly, competition can involve *adapted (programmed)* or *unadapted (incidental)* interactions. Adapted interactions often involve aggressive social behaviour, such as dominance hierarchies and territoriality, where only certain individuals high in the pecking order, or those holding territories, succeed in breeding. This is sometimes called *contest* competition because it involves aggressive contests between competing individuals. Unadapted competition, on the other hand, results from unintentional interaction between individual organisms utilizing the same resources and is sometimes called *scramble* competition because, in this case, everyone is involved in a mad, unorganized, scramble over the scarce resources.

Notice the parallels between competition and co-operation. Both can be induced by programmed (adapted) or incidental (unadapted) behaviour. Both are associated with the problem of obtaining resources, or avoiding being used as a resource by others. It should not be surprising that these two powerful principles have led to the parallel evolution of aggressive competitive behaviour and co-operative social behaviour in many species.

A major consequence of *intra*specific competition is that the survival and/or reproduction of the individual organism declines as the density of the population rises. (This is commonly called "density dependence" but, because this term has caused so much confusion and misunderstanding[12], we will try and avoid it in this book.) In other words, the action of the third principle causes the birth rate of individuals to fall and/or the death rate to rise as the population density increases, the net result being a decline in the realized per capita rate of change (Figure 5.1). In this case the feedback function has a negative slope and creates ‾feedback on population growth [remember condition (4.5b)].

Notice that the R-function in Figure 5.1 has two intercepts; A, the intercept with the ordinate being the maximum per capita rate of change possible in a given environment, and K, the intercept with the abscissa being the value of N at which $R = 0$. Because $R = 0$ when $N = K$, then K must be an equilibrium point. Notice that populations grow when they are below K, because $R > 0$ (births exceed deaths), and decline when above K (deaths exceed births or $R < 0$). This is indicated by arrows pointing towards the equilibrium point, K, which identify it as a *convergent* or *stabilizing* equilibrium, sometimes called an *attractor*.

5.1.1 Competition to avoid enemies

Although it is not as obvious as competition for scarce resources, organisms also compete in a sense to avoid being eaten by predators or infected by pathogens. This is sometimes called competition for "enemy-free space"[13]. One way of looking at this is to consider enemies as negative resources, in that the more of them there are, the worse off the individual. In contrast, the more positive resources like food and space there are, the better off the individual.

*Intra*specific competition to avoid enemies can also create ‾feedback and evoke the third principle if the risk of death or debilitation increases with population density. For example, when pathogenic micro-organisms are spread by contact between infected and uninfected hosts, the probability of infection usually increases with host density, and this can evoke

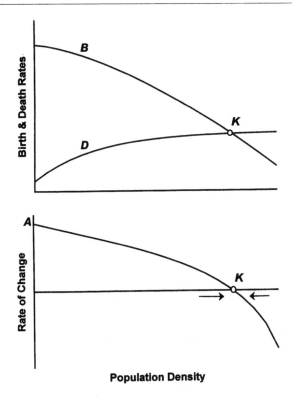

Population Density

Figure 5.1 Top, relationship between the per capita birth (B) and death (D) rates and population density (N_{t-1}) resulting from the effects of *intra*specific competition for scarce resources (the third principle); bottom, the corresponding relationship between the realized per capita rate of change, $R = \ln(1 + B - D)$, and population density – the R-function. Note that A is the maximum per capita rate of change and K is the equilibrium density; i.e. the point where $B = D$ and $R = 0$. The arrows signify the direction of population change in the neighbourhood of K.

the third principle and give rise to R-functions like that in Figure 5.1. Some predators can also evoke the third principle if they *switch* to more abundant prey species or *aggregate* in areas where prey are abundant, behaviours common to intelligent generalist predators such as birds and mammals, but also seen in some arthropods[14]. Under these conditions, being in a crowd is hazardous to the health of individuals because it is the crowd that draws the attention of the hungry predators.

Predators that switch to and aggregate on dense prey populations often have S-shaped or *sigmoid* behavioural responses[10] (Figure 5.2). In contrast to the cyrtoid behavioural response discussed in the last chapter

(Figure 4.2), the sigmoid response accelerates with prey density at first but then reverts to the typical cyrtoid form as the predator becomes satiated (Figure 5.2, top). The initial increase in the attack rate is due to switching and/or aggregation of predators on this particular prey species as its density rises relative to other alternative prey. As a consequence, competition for "enemy-free space" increases amongst members of the prey population, and this causes the per capita death rate of the prey to rise with prey density and the prey's R-function to have a negative slope (Figure 5.2). Notice, however, that the ‾feedback only operates at very low prey densities, and that ⁺feedback due to the second principle re-asserts itself as the prey population increases above a certain level. This change from ‾feedback to ⁺feedback produces what has been called a "predator pit"[15]. If sufficient predators are present in the environment, then the "pit" can be deep enough to produce an R-function with two equilibrium points, a low-density stabilizing equilibrium (J) and an unstable escape threshold (E) (Figure 5.2, bottom). In such cases, the predator can stabilize the prey population at a low-density equilibrium, J, but the prey can escape from control if its population exceeds the escape threshold, E. The occurrence of predator-regulated equilibria, and the degree of separation between stable and unstable equilibria ($E - J$), is critically dependent on the density of predators. If the predator popula-tion is reduced, say by habitat destruction, then the "pit" becomes shallower, and the separation smaller, until eventually the equilibrium points disappear altogether. When this happens the prey population can grow to very high densities; for example, an outbreak of prey. It is also evident that the larger the predator population, the deeper the "pit", the greater the separation between stable and unstable equilibria, and the lower the likelihood of prey escape. This, of course, has important impli-cations for pest managers.

5.2 MATHEMATICAL INTERPRETATION

Returning to the example of lions feeding on a herd of wildebeest, we now think about it from the point of view of the lion (the consumer) rather than that of the wildebeest (the prey). Each lion must eat a certain number of wildebeest, or wildebeest equivalents, each year in order to meet its energy requirements for survival, growth and reproduction. Call this the per capita *demand* for prey V, and the total demand of a pride of P lions is then VP. On the other hand, the total *supply* of prey, H, is given by the number of wildebeests, N, and other prey species, F, present in the environment; i.e. $H = N + F$. We would expect *intra*-specific competition amongst lions to be intense, and their R values low, when the total demand for prey is high relative to the supply. In other

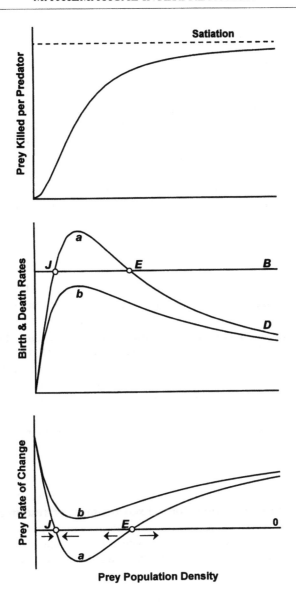

Figure 5.2 Top, the sigmoid behavioural response of a predator whose efficiency increases with prey density because of switching and/or aggregation. Middle, the resultant effect on the prey's death rate (D) with birth rate (B) assumed constant. Bottom, the corresponding R-function where J is a stabilizing low-density prey equilibrium enforced by predation and E is an unstable escape threshold; a identifies the prey death rate and R-function for a high density of predators and b for a low density of predators.

words, the intensity of *intra*specific competition amongst lions should be directly related, and R inversely related, to the *demand/supply* ratio. Once again there is an economic analogy in the price one has to pay for a good, price being directly related to demand/supply; i.e. the more people that want the good and the scarcer it is, the higher the price. The price that organisms pay is the increased energy spent in foraging when resources are scarce, a price that is also directly related to the demand/supply ratio. Taking this idea one step further, it is apparent that the higher the expenditure of energy by the consumer, the less is available for growth and reproduction and, therefore, the lower will be the realized per capita rate of change of the consumer, R_P. This inverse relationship between R_P and the demand/supply ratio can be expressed mathematically by

$$R_P = A_P - \frac{VP_{t-1}}{H},$$
(5.1)

where A_P is the maximum per capita rate of change of the consumer in a given environment when resources are superabundant (i.e. $H \to \infty$), and VP_{t-1}/H measures the reduction from this maximum due to *intra*specific competition for resources[16]. Notice that P_{t-1} identifies the consumer (predator) population as a dynamic variable while all the other letters represent constants. Also note that, if $N + F$ is substituted for H in equation (5.1), it is very similar to the model for *intra*specific co-operation (4.9), but the meaning of parameters W and V are, of course, not the same.

Because the supply of resources H is assumed to be constant in equation (5.1), it can be reduced to

$$R_P = A_P - C_P P_{t-1},$$
(5.2)

where $C_P = V/H$ represents the intensity of the "struggle for existence", or *intra*specific competition, for a fixed quantity of resources. Because equation (5.2) describes ̄feedback acting on the rate of population change, there will be a particular density K_P where $R_P = 0$ and the population is in equilibrium (K in Figure 5.1). Substituting K_P for P_{t-1} and 0 for R_P in equation (5.2) yields

$$0 = A_P - C_P K_P$$
$$K_P = A_P/C_P$$
(5.3)

Rearranging equation (5.2) as follows

$$R_P = A_P \left(1 - \frac{C_P P_{t-1}}{A_P} \right),$$
(5.4)

and substituting $1/K_P$ for C_P/A_P, gives

$$R_P = A_P \left(1 - \frac{P_{t-1}}{K_P} \right),$$ (5.5)

which is a discrete analogue of the famous *logistic equation* of population growth[17]. The equilibrium point of the logistic equation, K_P, is sometimes called the *carrying capacity* of the environment because it specifies the maximum population that can be sustained indefinitely in a particular environment[18]. It is because of the third principle, therefore, that populations of living organisms tend to reach equilibrium with the resources present in their environments. Note that the logistic equation applies strictly to a population of consumers utilizing a constant supply of resources.

It is also possible to include competition between different species (*inter*specific competition) in the mathematical model. For example, suppose the lions compete with leopards for wildebeest and other prey species, then equation (5.5) could be written

$$R_P = A_P \left(1 - \frac{P_{t-1} + L}{K_P} \right)$$ (5.6)

where L is the *relative* density of leopards; i.e. the density of leopards multiplied by their relative ability to compete with lions for the common prey species.

5.2.1 Non-linear *R*-functions

The logistic equation describes a linear relationship between the realized rate of change of an average individual and the density of the population in which it resides (it is also a linear function of the consumer/resource ratio). The logistic is, therefore, the simplest *R*-function for a population of consumers obeying the third principle. There are many situations, however, when the relationship is not expected to be linear as, for example, when social behaviour determines the outcome of the interaction (i.e. adapted *intra*specific competition).

The problem of non-linearity in the *R*-function can be accommodated by the addition of another parameter[19], the *coefficient of curvature*, Q_P,

$$R_P = A_P \left[1 - \left(\frac{P_{t-1}}{K_P} \right)^{Q_P} \right],$$ (5.7)

or

$$R_P = A_P - C_P P_{t-1}^{Q_P}.$$ (5.8)

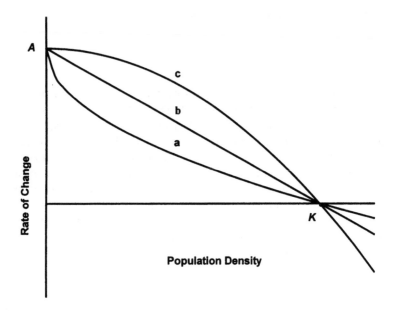

Figure 5.3 Effect of the coefficient of curvature on the relationships between the realized per capita rate of change, $R = \ln(1 + B - D)$, and population density, P_{t-1}, as in equation (5.7) when $A_P = 1$, $K_P = 1000$, and $Q_P = 0.5$ (*a*), 1.0 (*b*) and 2.0 (*c*).

When $Q_P = 1$ the R-function is linear, when $Q_P < 1$ it is concave (the slope of the function decreases with density), and when $Q_P > 1$ it is convex (the slope becomes steeper as density rises) (Figure 5.3). In situations of adapted *intra*specific competition, as in territorial animals for example, Q_P is expected to be greater than unity because competition should become stronger as the carrying capacity is approached and most of the territories occupied.

5.3 POPULATION DYNAMICS UNDER THE THIRD PRINCIPLE

5.3.1 Cellular automata

The dynamic consequences of the third principle can be illustrated in populations of artificial "organisms" called *cellular automata*[20]. Cellular automata "live" on a square lattice composed of a finite number of cells, something like a chess board. On this lattice an "organism" can have up to eight neighbours, with the actual number of neighbours determining whether an automaton will reproduce or die. For example, let

1. automata with more than five neighbours die from overcrowding (third principle);
2. those with less than two neighbours die from isolation (second principle);
3. those with two to five neighbours reproduce offspring into adjacent empty cells (first principle); and
4. a certain proportion of the offspring be killed by random disturbances (exogenous effect).

Beginning with a few automata positioned somewhere on the board, each individual is checked in turn to see how many neighbours it has. After all automata have been checked, those with more than five or less than two are eliminated (rules 1 and 2) while those with two to five neighbours reproduce offspring into adjacent empty cells (rule 3). A random number between zero and one is then chosen and, if that number is less than the probability of random death, then the newborn automaton is eliminated (rule 4). Finally, the number of surviving "organisms" is plotted to show the dynamics of the population through time (Figure 5.4, left). Notice how the population grows rapidly at first (+feedback due to the action of the first and second principles) but then levels off as the lattice becomes fully occupied (−feedback due to the third principle). Also notice that the growth curve is S-shaped or sigmoid. The levelling off of the growth curve around the "carrying capacity" of the board is due to increasing mortality

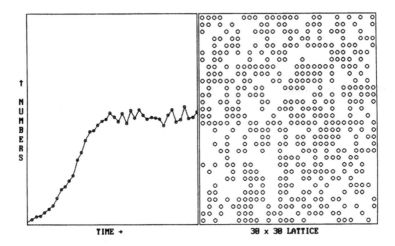

Figure 5.4 The dynamics of 40 "generations" of cellular automata inhabiting a 30 × 30 lattice using the rules specified in the text and a random probability of juvenile death = 0.4. The graph on the left shows the time trajectory and that on the right the spatial distribution of automata at the end of the simulation.

from overcrowding as the population expands to occupy all the available space; i.e. space is the resource being competed for.

5.3.2 Numerical simulations

Of course the dynamics can also be investigated by numerical solution of the mathematical models developed earlier. For example, set the parameters of equation (5.7) to $A_P = 0.8$, $K_P = 1000$, $Q_P = 1$, and the initial population density $P_0 = 10$, and then calculate the realized per capita rate of change

$$R_P = A_P [1 - (P_0/K_P)^{Q_P}] = 0.8(1 - 10/1000) = 0.79.$$

Substitute this value in the forecasting equation (3.12) to give the number of organisms in the following year

$$P_1 = P_0 e^R = 10e^{0.79} = 22.$$

Continuing,

$$R_P = 0.8(1 - 22 / 1000) = 0.78$$
$$P_2 = 22e^{0.78} = 48$$

and so on (Figure 5.5, top left). Notice how the population grows to equilibrium K_P along a *sigmoid* or S-shaped trajectory, much like the cellular automata (Figure 5.4), and that it approaches equilibrium smoothly, without oscillation. We call this a *smooth* or *asymptotic* approach to equilibrium.

We can now change the value of the parameters of the R-function to see how this affects the dynamics of populations obeying the third principle. This is known as *sensitivity analysis* because one can observe how sensitive the model is to variations of its parameters. For example, the sensitivity of the linear logistic model to changes in the maximum per capita rate of change, A_P, can be examined by varying this parameter while all other parameters are kept constant. Such an analysis will show that the model exhibits (Figure 5.5):

1. *Smooth* growth to equilibrium when $A_P \leq 1$.
2. *Damped oscillations* around equilibrium when $1 < A_P < 2$.
3. *Periodic oscillations* that become more and more complex when $A_P \geq 2$, eventually leading to aperiodic *chaos*[21] when $A_P > 2.692$. In these cases, the equilibrium point is locally unstable but the oscillations persist indefinitely. Notice that chaotic fluctuations are aperiodic because their amplitude, the difference between peaks and troughs, varies in an irregular manner.

The most obvious result of the sensitivity analysis is that the dynamics become less stable and more complex as the value of the parameter A_P

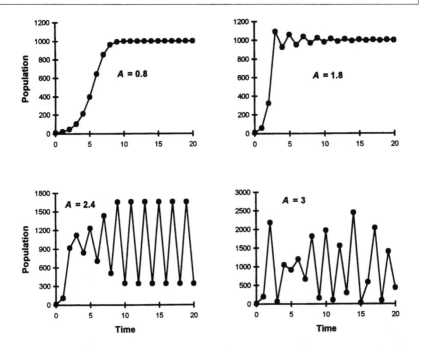

Figure 5.5 Sensitivity of the logistic R-function to variations in the maximum per capita rate of change A_P given that $K_P = 1000$, $Q_P = 1$, and $P_0 = 10$.

gets larger. The reason for this is that A_P determines the rate of approach to equilibrium and, the faster the approach, the further the population will overshoot its equilibrium level. Large overshoots of equilibrium result in stronger ⁻feedback, correspondingly large undershoots of equilibrium and, consequently, more violent fluctuations. It is particularly important to notice that the oscillations have a characteristic sharp, or *saw-toothed*, pattern that repeats itself every second time step (period-2 oscillations). High-frequency fluctuations with these characteristics are typical of *first-order* dynamic systems (more about this in Chapter 6). The general result of this analysis is that ⁻feedback due to the action of the third principle tends to stabilize populations at or near a characteristic density or carrying capacity, but that the actual stability of the equilibrium point depends on the values of certain parameters. In this sense ⁻feedback is a *necessary* but not *sufficient* condition for stable population dynamics.

Let us now examine the sensitivity of the logistic equation to changes in the coefficient of curvature (Figure 5.6). Notice that concave R-functions ($Q_P < 1$) are more stable than convex ones ($Q_P > 1$), other things being equal. In general, variations in this parameter have similar effects to A_P, with larger values producing more complex and less stable

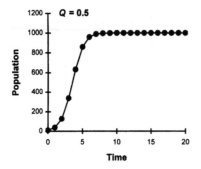 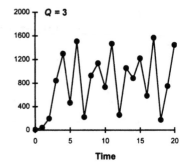

Figure 5.6 Sensitivity of the logistic equation to the coefficient of curvature, Q_P, when the constant parameters are set at $A_P = 1.5$, $K_P = 1000$, and $N_0 = 10$.

dynamics. The reason for this is that the stability of the equilibrium point is determined by the slope of the R-function in the vicinity of equilibrium; the steeper the slope, the less stable the equilibrium. As both A_P and Q_P directly affect the slope of the R-function near equilibrium, they also determine the stability of the equilibrium point.

Similar analysis with the parameter K_P will show that it has no effect on the dynamic behaviour of the model but, of course, does affect the level at which the population attains equilibrium.

5.3.3 Random variability

It is commonly assumed that oscillations observed in natural populations are due to random exogenous variability. However, it is clear from the sensitivity analysis that regular, and even irregular fluctuations can be produced by discrete population models when environmental conditions are constant (Figures 5.5 and 5.6). How then does environmental variability affect the dynamics of populations obeying the third principle?

The dynamics of populations living in a noisy world is simulated, as before, by using the stochastic version of the forecasting equation (3.13). The first thing to notice is that environmental variability can cause a previously stable trajectory (as in Figure 5.5, top left) to fluctuate irregularly around its equilibrium level (Figure 5.7), and that the amplitude of oscillations depends on the amount of noise.

The second thing to notice is that environmental variability causes trajectories that would normally damp to a stable equilibrium point (as in Figure 5.5, top right) to continue oscillating with amplitude determined by the amount of noise (Figure 5.8).

Finally, populations exhibiting strong oscillatory tendencies in their undisturbed dynamics (as in Figure 5.5, bottom) seem to be less sensitive

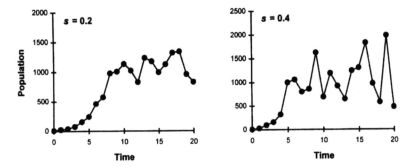

Figure 5.7 The dynamics of an asymptotically stable population (see Figure 5.5, top-left; A_P = 0.8, K_P = 1000, Q_P = 1) when subjected to small (s = 0.2) and large (s = 0.4) random disturbances.

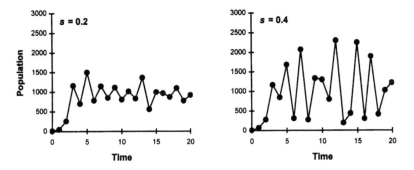

Figure 5.8 The dynamics of a damped-stable population (see Figure 5.5, top-right; A_P = 1.8, K_P = 1000, Q_P = 1) when subjected to small (s = 0.2) and large (s = 0.4) random disturbances.

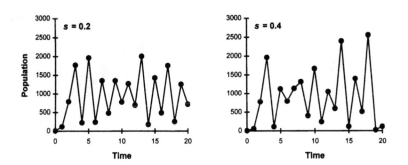

Figure 5.9 The dynamics of a population with an unstable equilibrium (see Figure 5.5, bottom-left; A_P = 2.4, K_P = 1000, Q_P = 1) when subjected to small (s = 0.2) and large (s = 0.4) random disturbances.

to random disturbances (Figure 5.9) than more stable populations (compare with Figures 5.7 and 5.8). Notice also that it may be difficult to distinguish between periodic trajectories with a small amount of noise (Figure 5.9, left), chaotic trajectories without noise (Figure 5.5, bottom right), and a damped-stable population with a lot of noise (Figure 5.8, right).

5.3.4 Phase portraits

At this point it is instructive to introduce the concept of *phase space*, the *n*-dimensional co-ordinate system that describes all possible states of a system composed of *n* dynamic variables. In our case there are only two dynamic variables, population density and the realized per capita rate of change, so only two dimensions are needed. Hence, phase space is defined by the graph of R plotted against P (or N). Within this phase space, the dynamics of the system are depicted by the trajectory of co-ordinates (P, R) joined by lines called *vectors*. When all vectors of a dynamic trajectory are plotted in phase space, a *phase portrait* is obtained. For example, the phase portrait for one of the earlier simulations is shown in Figure 5.10. Notice how all the vectors point towards, or pass close to, the equilibrium point, K. This is the typical phase portrait created by the action of the third principle, or any *first-order* discrete dynamic process.

5.3.5 Environmental forcing

In addition to random variation, populations can also be affected by more predictable environmental changes, or what is called *environmental*

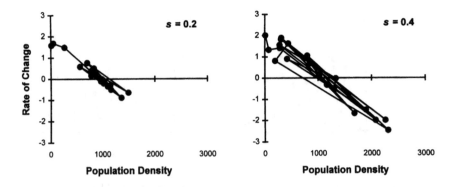

Figure 5.10 Vectors of population change in *P-R* phase space, or the phase portrait of the dynamic trajectories (numbers from Figure 5.8).

forcing. For example, environmental forcing can act on the maximum per capita rate of change, *A*, because this parameter reflects the basic suitability of the environment in the absence of *intra*specific competition. Hence, the value of *A* can be considered a function of prevailing environmental conditions, including such things as temperature, rainfall, topography, soils, etc. Similar arguments can be made for the parameters *C*, *K* or *H*. In particular, these parameters must be affected by the abundance of resources like food, water, nesting places, and so on.

When environmental conditions change gradually they can induce *trends* in the dynamic variables. For instance, global warming caused by gradual and persistent increases in atmospheric carbon dioxide and other "greenhouse" gases could cause slow changes in the reproductive potential, survival rate, or carrying capacity of certain species. Consistent changes in the values of these parameters can be simulated by a linear function

$$X_t = X_0 + b(t - t_0).\qquad(5.9)$$

where X_t is the value of the parameter at time *t*, X_0 is its original value at time t_0, and *b* is a time-dependent forcing constant. The effects of trends in the maximum per capita rate of change, *A*, and the carrying capacity, *K*, are simulated in Figure 5.11. Not surprisingly, gradual environmental forcing of the maximum per capita rate of change alters the stability of the equilibrium and the degree of fluctuation around equilibrium (Figure 5.11, left), while forcing of the carrying capacity has no effect on stability, but causes the mean density to change in a systematic manner (Figure 5.11, right).

Environmental forcing can also cause sudden discontinuities or *shifts* in parameter values. For example, a volcanic eruption may change the environment from very favourable to very unfavourable in a matter of hours.

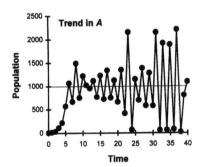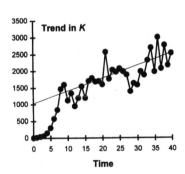

Figure 5.11 Simulation of linear trends in the maximum per capita rate of change *A* when the carrying capacity *K* is held constant (left), and the carrying capacity *K* while *A* remains constant (right).

Discontinuous changes in parameter values can be simulated by a step function such as

$$X_t = \begin{cases} X_0, t \leq t_{step} \\ X_{new}, t > t_{step} \end{cases},$$ (5.10)

where X_{new} is the new value of the parameter which replaces the old one X_0 at time $t = t_{step}$. The effects of discontinuous shifts in the maximum per capita rate of change and the carrying capacity are simulated in Figure 5.12. Notice the sudden changes in stability and the average density. We will see examples of similar dynamics in real populations later in this book.

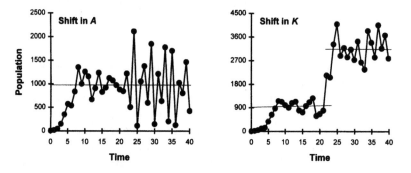

Figure 5.12 Simulation of discontinuous shifts in the per capita rate of change A (left) and carrying capacity K (right).

5.4 SUMMARY

In this chapter we

1. Defined the third principle of population dynamics, which recognizes that individual organisms may compete for resources when their populations become large.
2. Defined the concept of competition and the notions of adapted (programmed) and unadapted (incidental) *intra-* and *inter*specific competition.
3. Derived the R-function for the relationship between population density and the realized per capita rate of change under the third principle.
4. Derived a mathematical equation for the third principle – the logistic equation – for consumers foraging for fixed resources.
5. Generalized the logistic equation to include *inter*specific competition and non-linear competitive interactions.

6. Demonstrated the dynamic consequences of the third principle in simple artificial-life systems (cellular automata) and by simulation with the models developed in this chapter.
7. Analysed the sensitivity of the non-linear R-function to variations of its parameters, and showed that stability was affected by the parameters A and Q but not K.
8. Showed how random environmental disturbances caused persistent saw-toothed oscillations in populations that would otherwise attain a stable equilibrium, and increased the amplitude of population fluctuations.
9. Introduced the concepts of phase space and the phase portrait.
10. Demonstrated how gradual or sudden changes in the external environment (forcing) can affect the dynamics of populations obeying the third principle (trends and shifts).

5.5 EXERCISES

1. Simulate the dynamics* of a population governed by the third principle when the initial density is 10 individuals, the maximum per capita rate of change $A = 0.8$, the coefficient of *intra*specific competition $C = 0.001$, the coefficient of curvature $Q = 1$, and the run length is 20 time steps. Plot the trajectory. Calculate the density of the population at carrying capacity?
2. Repeat the above simulation* but this time add some noise ($s = 0.2$). Plot the trajectory and also the vectors in (N, R) phase space.
3. Repeat the two simulations* above with $A = 1.8$ and $A = 3$.
4. If you have the *PAS* program *PG2*, use it to simulate the growth dynamics of a population of cellular automata in constant and variable environments. One can also program a computer to do these simulations.

*These problems can be done by hand, on a typical computer spreadsheet, or using the *PAS* software program *P1b* (see Preface for details).

6 | The fourth principle: circular causality

Groups (of organisms) may be acted upon by their environment, and they may react upon it. If a set of properties in either system changes in such a way that the action of the first system on the second changes, this may cause changes in the properties of the second system which alter the mode of action of the second system on the first. Circular causal paths can be established in this manner. (G. E. Hutchinson, 1948)

When discussing the first three principles of population dynamics we assumed that the environment affected, but was not affected by, the population in question. In other words, environmental factors acted as constants or as one-way *exogenous* inputs (Figure 1.2, top). Thus, although population density influenced the degree of competition and co-operation between individuals, it had no effect on the environment and, in particular, on the abundance of resources or enemies. The fourth principle recognizes that populations can affect the properties of their environments, and that this can create circular causal pathways linking populations to their resources, enemies, or other environmental components[22] (Figure 1.2, bottom). Think of it this way: if a population increases, more food is available for its natural enemies, which then increase, eat more prey, and thereby reduce the density of the prey population. This is a circular causal process because N (the prey population) affects P (the enemy population) which in turn affects N. In other words, N affects itself, after a period of time, through P. This particular circular causal process creates a *second-order* ¯feedback effect on both predator and prey populations; i.e. N has a positive effect on P and P has a negative effect on N, and the product of a positive and a negative effect is a negative loop. As we will see in the next chapter, circular causal processes often create time delays in the feedback structure, which can have important effects on the dynamics of the system. Of course, it is also possible for circular causal processes to have a net ⁺feedback effect.

6.1 ENVIRONMENTAL FACTORS

Circular causality involves the interaction between a population and its environment, where the environment consists of all the physical and biological factors occupying the space within which the population exists. Environments contain both the resources that organisms need to survive and reproduce (positive resources) and the enemies and other dangerous elements that make survival and reproduction difficult (negative resources). In this book we will distinguish between two basic kinds of environmental factors as discussed below.

6.1.1 Non-reactive factors

Environmental factors whose abundance, concentration, or properties *do not change* in response to the density of the population will be known as passive or non-reactive environmental factors. Non-reactive factors act as one-way exogenous effects (Figure 1.2, top) and are not involved in circular causality. They are often physical factors like climate, solar energy, mineral nutrients, and so on, but as we will see later, biotic factors can also be non-reactive. Space is a major passive factor in that organisms need space in which to gather sunlight, chemicals, food and water, or in which to avoid or escape enemies (enemy-free space), but the total space available does not change in response to population density. In fact, almost all non-reactive factors can be considered, in one way or another, to involve space.

Because passive factors do not create circular causal pathways, they do not evoke the fourth principle. Hence, the dynamics of populations interacting with passive environmental factors are determined by the first three principles and the endogenous system is composed of a single dynamic variable, the density of the population, N. Because of this, the feedback resulting from interactions between populations and non-reactive factors is depicted by an arrow from the population back to itself (Figure 1.2). This type of feedback is sometimes termed self-regulation (or self-loops) and, as mentioned previously, it gives rise to *first-order* dynamics. Note that the self-regulation arrow can have a positive or a negative sign depending on whether the second or third principles are operating at the time.

It is important to realize that, although the total quantity of a non-reactive factor present in the environment does not change in response to population density, its physical location in the environment can change.

For example, phosphorus is extracted from the soil by plants and stored in plant tissues for decades, but this same phosphorus is released back into the soil upon the death and decay of the plant. The total quantity present in the environment, including soil, plants and animals, remains roughly the same. Similarly, predators may aggregate in certain locations, say where prey are more abundant, yet the density of the predator population remains unchanged over larger areas. Such predator populations are also considered non-reactive.

6.1.2 Reactive factors

Environmental factors are classified as reactive if they change quantitatively (and occasionally qualitatively) in response to changes in the density of the population, and if that change is sufficient to affect the birth and death rates of individuals in the population. Reactive factors are usually (but not always) living organisms that are used as food, or enemies that use the subject population for food (Figure 1.2, bottom). Of course, populations of food organisms change quantitatively because they are consumed and, thereby, removed permanently from the environment. Enemy populations also change quantitatively because they reproduce more offspring when prey are abundant. If these changes in the reactive factors are sufficient to induce changes in the per capita birth or death rates of the subject population, then a circular causal pathway from the population back to itself is created and the fourth principle is evoked.

It is also possible, but not common, for physical factors to be affected by population density and then to take part in circular causal pathways. For example, air and water pollution caused by expanding human populations may feed back to affect human mortality from cancer and other diseases. The dynamic consequences of such circular causal pathways will become evident later on.

6.2 MATHEMATICAL INTERPRETATION

Let us begin with the simple linear R-function describing the "struggle for existence" for fixed resources amongst a population of consumers, P [equation (5.2)]

$$R_P = A_P - C_P P_{t-1}, \tag{6.1}$$

where A_P is the maximum per capita rate of change of the consumer population, P_{t-1} is the density of the population, and C_P measures the degree of *intra*specific competition for fixed resources. Because this equation describes the situation where resources are not affected by population density, it applies only to non-reactive environmental factors.

Equation (6.1) can be generalized to include interactions with reactive factors by considering the circular causal pathway between consumer and resource populations. In this case, the density of each population not only affects its own per capita rate of change but also affects that of the other species. Now if equation (6.1) describes the direct feedback (the self-loop) in the consumer R-function, then to complete the model we need an expression for the effect of a reactive resource, N_t, on the R-function of its consumer, P_t. Remembering that equation (6.1) can also be expressed as

$$R_P = A_P - \frac{VP_{t-1}}{H},$$ (6.2)

and that H can be interpreted as $N_{t-1} + F$, where F is a fixed quantity of alternative prey, then equation (6.2) can also be written in terms of its reactive resource, N_{t-1},

$$R_P = A_P - \frac{VP_{t-1}}{N_{t-1} + F}.$$ (6.3)

Remember that the time subscript $t - 1$ identifies the resource population as a dynamic variable. Finally, equations (6.1) and (6.3) can be integrated into a single model for a consumer utilizing both passive and reactive resources

$$R_P = A_P - C_P P_{t-1} - C_{PN} \frac{P_{t-1}}{N_{t-1} + F}.$$ (6.4)

For consistency, the per capita consumer demand for resources, V, has been replaced by the coefficient of *intra*specific competition amongst consumers for reactive resources, C_{PN} . Of course, these parameters are directly related because the larger the demand for resources, the greater the *intra*specific competition for those resources, all else being equal. The equation is composed of the following parts:

1. A_P, the maximum per capita rate of change of the consumer population when resources are unlimited (no *intra*specific competition for fixed or reactive resources).
2. $C_P P_{t-1}$, the reduction from the maximum rate of change due to *intra*specific competition for non-reactive resources (usually space). This self-regulating effect manifests itself as a first-order feedback loop from the consumer population back to itself through its per capita rate of change.
3. $C_{NP} P_{t-1}/(N_{t-1} + F)$, the reduction from the maximum rate of change due to *intra*specific competition for reactive resources (usually food). This is sometimes referred to as a "bottom-up" effect because the dynamics of the consumer population are affected by the lower trophic level.

Equation (6.4) defines the R-function for a consumer population, P, utilizing a reactive resource population, N. In order to complete the mutually causal loop, we also need an R-function for the resource (prey) population. Assuming the prey population is regulated by *intra*specific competition for non-reactive resources when no predators are present in the environment, then the R-function for the prey in the absence of predators is also given by the logistic equation

$$R_N = A_N - C_N N_{t-1}.$$

In the presence of predators, however, the per capita rate of change of the prey must be reduced by the per capita death rate due to predation; i.e. equation (4.7). Hence, the prey R-function becomes

$$R_N = A_N - C_N N_{t-1} - C_{NP} \frac{P_{t-1}}{N_{t-1} + F}, \tag{6.5}$$

where, for consistency, the coefficient of *intra*specific competition for "enemy-free space", C_{NP}, replaces the impact of the predator, W, in equation (4.7). The prey R-function contains the following components:

1. A_N, the maximum per capita rate of change of the prey population when resources are unlimited (no *intra*specific competition for fixed resources) and predators are absent from the environment.
2. $C_N N_{t-1}$, the reduction from this maximum due to *intra*specific competition for fixed resources. This self-regulating effect manifests itself as ⁻feedback from the prey population to itself through its own per capita rate of change.
3. $C_{NP} P_{t-1}/(N_{t-1} + F)$, the reduction from this maximum due to predation. This effect is sometimes referred to as the "top-down" effect because it represents the effect on the population by the higher trophic level.

Equations (6.4) and (6.5) describe the R-functions for reactive populations of predators and prey, or consumers and resources. Note that both equations are very similar in that the per capita rates of change are linear functions of the same predator/prey ratio, $P_{t-1}/(N_{t-1} + F)$. Equations with these properties are sometimes called *ratio-dependent*[23]. When equations (6.4) and (6.5) are linked together they create a mutually causal *second-order*[24] ⁻feedback loop; i.e. the loop $N \rightarrow P \rightarrow N = N \rightarrow N$ (Figure 1.2, bottom). This ⁻feedback loop describes the reproductive response of the predator population to the density of its prey and the subsequent suppression of the prey population by predation. The increase in predator reproduction in response to increasing prey density is often called the *numerical response* of the predator to the density of its prey.

6.3 DYNAMICS UNDER THE FOURTH PRINCIPLE

6.3.1 Cellular automata

The dynamic consequences of the fourth principle can be illustrated by systems of prey and predator automata. Remember that cellular automata are artificial "organisms" whose survival and reproduction depends on how many neighbours they have. We now recognize two types of automata, prey and predator, whose survival and reproduction is governed by the following rules:

1. Predator automata eat any prey occupying adjacent cells.
2. Predators that eat no prey in a "generation" die of starvation.
3. Predators that eat more than one prey reproduce offspring into the cells that contained all but the first prey eaten (i.e. one prey is required to sustain life and all others are converted into predator offspring).
4. Surviving prey reproduce into any empty adjacent cells.
5. A random proportion of newborn predator and prey automata die.

When this predator–prey game is simulated on a computer, cycles of abundance of predator and prey are frequently observed (Figure 6.1, left). The prey population increases at first but is soon followed by an increase of predators, which then suppress the prey population and later crash themselves for lack of food. The spatial dynamics of prey and predator

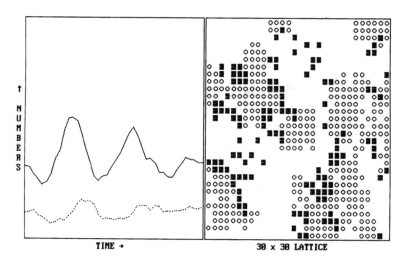

Figure 6.1 The results of a simulation with prey and predator automata feeding and reproducing on a 30 × 30 lattice (prey are represented by open circles, predators by black squares).

automata are also quite interesting. Predators form into aggregations that eat holes in the prey population and then die out in the empty spaces. When seen on a computer simulation, the predators seem to be chasing the prey around in the available space and, in so doing, creating complex shifting patterns that are reminiscent of the spatial complexity in the real world (Figure 6.1, right).

6.3.2 Numerical simulation

The effect of the fourth principle can also be simulated by numerically solving the coupled predator–prey equations (6.4) and (6.5). Starting with given densities of predators and prey (P_0, N_0), the realized per capita rate of change of the predator population is calculated with equation (6.4) and the prey population with equation (6.5). Then the new densities of both species are calculated with the familiar step-ahead forecasting rule [equation (3.12)]. This process is repeated for as many time steps as desired (Figure 6.2).

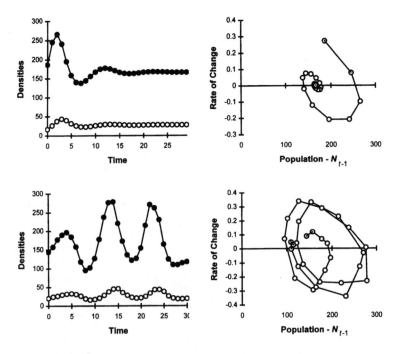

Figure 6.2 Left, the dynamics of a coupled predator (solid circle)–prey (open circle) interaction simulated with equations (6.4), (6.5) and (3.12) with parameters $A_P = A_N = 1$, $C_P = 0$, $C_N = 0.002$, $C_{PN} = 6$, $C_{NP} = 4$, $F = 0$ in a constant (top) and a randomly varying (bottom; $s_P = 0$, $s_N = 0.1$) environment; right, the corresponding phase plots for the prey population.

The most significant thing that emerges from these simulations is that, for certain values of the parameters, the two populations oscillate with fairly long period. These low-frequency cycles are quite unlike the sharp, high-frequency or saw-toothed, oscillations seen before (Chapter 5). Low-frequency oscillations such as these are often called "population cycles" and so we will use the term *cyclical* oscillations to describe them. It is important to note that the predator–prey oscillations may damp to a stable equilibrium in an invariant environment (Figure 6.2, top), but that the cyclic dynamics are amplified and sustained by a small amount of random variability (Figure 6.2, bottom). It is also important to note that, if the vectors are plotted in phase space, they orbit around the equilibrium rather than pointing towards it as in first-order dynamic systems (Figure 6.2, right). Broad circular phase portraits like this are typical of *higher order* dynamics generated by the action of the fourth principle. In the present case, of course, where there are two interacting species (or two equations), the resulting dynamics are second order.

6.3.3 Equilibrium isoclines

At this point it is useful to introduce the concept of equilibrium *isoclines* or *nullclines*. Equilibrium isoclines define a set of points in phase space on which the growth rates of the populations in question are zero. For example, we know that a population will be in equilibrium when its per capita rate of change is zero because, under this condition, births and deaths are equal (Chapter 4). Hence, the equilibrium *isoclines* for a prey or a predator population can be found by solving equations (6.4) and (6.5)

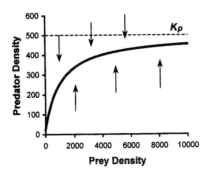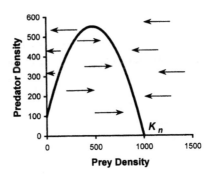

Figure 6.3 Prey–predator (N, P) phase space showing the right-slanting predator equilibrium isocline calculated from equation (6.4) with parameters $A_P = 1$, $C_P = 0.002$, $C_{PN} = 2$, $F = 50$ (left), and the parabolic prey isocline calculated from equation (6.5) with parameters $A_N = 2$, $C_N = 0.002$, $C_{NP} = 1$ and $F = 50$ (right). The arrows show directional vectors of population change in phase space, and $K_P = A_P/C_P = 500$ and $K_N = A_N/C_N = 1000$ are the carrying capacities of predator and prey populations, respectively.

when $R_P = R_N = 0$. Starting with the predator equation (6.4), and omitting time subscripts for simplicity, we have

$$0 = A_P - C_P P - C_{PN} P / (N + F)$$

$$P = \frac{A_P}{C_P + \dfrac{C_{PN}}{N + F}} \tag{6.6}$$

This isocline originates at the point $A_P/(C_P + C_{PN}/F)$ on the predator axis [i.e. solve equation (6.6) for P when $N = 0$] and gradually approaches a maximum at the carrying capacity of the predator, $K_P = A_P/C_P$ [i.e. solve equation (6.6) for P as $N \rightarrow \infty$] (Figure 6.3, left). Notice that the vectors of predator population change appear as vertical arrows converging on the zero-growth isocline. From this we see that the predator population grows when it is below its isocline and declines when it is above it.

The prey isocline is found by solving equation (6.5) for $R_N = 0$

$$0 = A_N - C_N N - C_{NP} P / (N + F)$$

$$P = \frac{N(A_N - C_N N) + F(A_N - C_N N)}{C_{NP}} \tag{6.7}$$

This isocline has a humped or parabolic shape that begins at the point FA_N/C_{NP} on the predator axis [i.e. solve equation (6.7) for P when $N = 0$] and intercepts the prey axis at the "carrying capacity" of the prey population, $K_N = A_N/C_N$ [i.e. solve equation (6.7) for N when $P = 0$] (Figure 6.3, right). Notice that the vectors of prey population change appear as horizontal arrows diverging from the ascending left-hand arm of the isocline and converging on the descending right-hand arm. As with the predator, a prey population grows when it is below its isocline and declines when it is above it. However, while the predator isocline is potentially stable along its entire length, the prey isocline is only stable along its descending arm. In both cases, stability is induced by the action of the third principle, *intra*specific competition for resources. On the other hand, the left-hand arm of the prey isocline is unstable because of the influence of the second principle, incidental *intra*specific co-operation in defence against predator attack.

When the two isoclines are superimposed in phase space, a *community equilibrium* occurs at the intersection because, at this point, both populations remain constant (Figure 6.4). Notice that predator–prey phase space is divided into four regions by the equilibrium isoclines:

I.   In this region both populations grow because they are both below their respective isoclines, so that all vectors of population change point towards the upper-right corner.

II. In this region the predator population grows (below its isocline) while the prey declines (above its isocline), so that all vectors of population change point towards the upper-left corner.

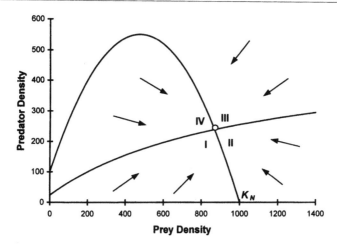

Figure 6.4 Prey–predator (N, P) phase space showing the superimposition of predator (curve sloping to the right) and prey (parabolic curve) equilibrium isoclines calculated in Figure 6.3. Arrows show directional vectors resulting from simultaneous changes in prey and predator populations (resultant vectors) in the four regions of phase space defined by the isoclines. K_N is the carrying capacity for the prey population in the absence of predation.

III. In this region both populations decline because they are above their respective isoclines, and all vectors point towards the lower-left corner.

IV. In this region the prey population grows (below its isocline) while the predator declines (above its isocline), and all vectors point towards the lower-right corner.

Notice that all the vectors of population change point towards the equilibrium point, in much the same way as do the vectors of a first-order dynamic process. In fact, as the community equilibrium moves to the right, down the stable descending arm of the prey isocline, the dynamics become more and more dominated by the high-frequency, or saw-toothed, oscillations characteristic of first-order feedback (Figure 6.5, top). This is because the prey population becomes more and more influenced by *intra*specific competition for passive resources as it gets closer to its carrying capacity, K_N. Under these conditions the predator has little impact on the dynamics of the prey population, which acts more or less as a passive resource, and the predator is limited by *intra*specific competition for prey. In other words, the third principle dominates the dynamics of both populations. This kind of regulation of food chain dynamics from the bottom of the chain towards the top is called "bottom-up" or "donor controlled" dynamics.

In contrast, as the community equilibrium moves further to the left, the dynamics become less stable and more cyclical (Figure 6.5) because of

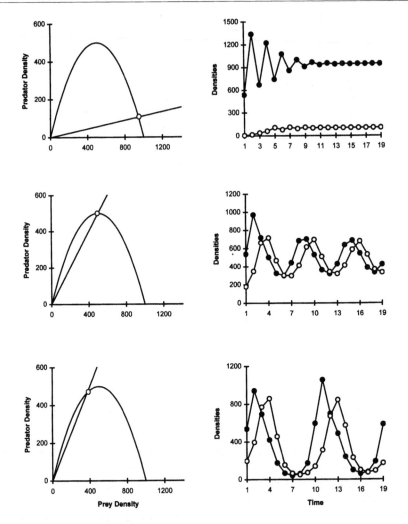

Figure 6.5 Equilibrium isoclines (left) for predator (straight line) and prey (parabola), and time trajectories (right) calculated from equations (6.4) and (6.5) with parameters $A_N = 2$, $C_N = 0.002$, $C_{NP} = 1$, $A_P = 1$, $C_P = 0$, $F = 0$ set constant, and C_{PN} variable; $C_{PN} = 9$ (top), $C_{PN} = 1$ (middle), and $C_{PN} = 0.8$ (bottom).

the influence of the second principle (*intra*specific co-operation to avoid being eaten) and the fourth principle (mutual causality between predator and prey populations). This kind of food chain regulation by higher trophic levels is called "top-down" or "recipient controlled" dynamics.

As we have seen, the dynamics of predator–prey interactions are extremely sensitive to the location of the community equilibrium. In turn, the location and stability of the community equilibrium is sensitive to the parameters of the predator–prey model in the following way:

1. Increasing the prey's maximum per capita rate of change, A_N, moves the equilibrium point to the right and reduces the period of oscillation. However, the equilibrium point is destabilized by large values of A_N because of first-order instability (i.e. the pathway to chaos illustrated by Figure 5.5). Stability is also decreased when A_N becomes small because the predator can then drive the prey to extinction. Thus, the system is most stable for intermediate values of the prey's maximum per capita rate of change.

2. Increasing the maximum per capita rate of change of the predator, A_P, moves the equilibrium point to the left, increasing the period of oscillation and decreasing stability. This is because predators with high reproductive potentials have greater impact on the dynamics of their prey and greater top-down control of the prey population. Predators with very high maximum per capita rates of change, relative to those of their prey, can drive the prey to extinction.

3. Increasing the carrying capacity for the prey, K_N, or decreasing *intra*specific competition for fixed resources, C_N, has no effect on stability but increases the equilibrium density of both species.

4. Increasing the carrying capacity for the predator, K_P, or decreasing *intra*specific competition for fixed resources, C_P, moves the equilibrium left, decreasing stability and increasing the period of oscillation.

5. Increasing *intra*specific competition for "enemy-free space", C_{NP}, or the impact of the predator, W, moves the equilibrium point left and decreases stability.

6. Increasing the intensity of *intra*specific competition amongst predators for prey, C_{PN}, or the demand for prey, V, moves the community equilibrium right and increases stability (Figure 6.5).

7. Increasing the alternative food for the predator, F, moves the equilibrium right, increases stability, and raises the intercept of the isoclines on the predator axis (c.f. Figure 6.4 when $F = 50$ with Figure 6.5 when $F = 0$).

It is important to understand that predator–prey interactions can result in a continuum of dynamic behaviours, from high-frequency saw-toothed oscillations, reminiscent of first-order dynamics (Figure 6.5, top), to low-frequency cyclical oscillations due to second-order feedback (Figure 6.5, middle and bottom). This is because cyclic dynamics are only observed if second-order ¯feedback between predator and prey populations is stronger than first-order ¯feedback due to *intra*specific competition within prey and/or predator populations. If the impact of the predator is small (C_{NP} small), or if prey have a very high maximum per capita rate of change relative to that of the predator (A_N large or A_P small), or if predators compete strongly for resources (large C_P or C_{PN}), then the feedback between predator and prey populations is weak, and the fourth principle is less likely to dominate the feedback structure. Under these conditions the dynamics of the prey population are controlled by first-order feedback due to intraspecific competition for fixed resources, and the predator population will merely track the saw-toothed oscillations of its prey (= bottom-up control). The fourth principle is more likely to dominate (= top-down control) if prey are consumed more rapidly than they can reproduce *and* if the predator reproduces sufficient offspring and/or has sufficient impact to suppress the prey population; i.e. the predator has a strong *numerical response.* This is more likely to happen if predators have large appetites, are efficient at locating prey when they are sparse, and/or specialize on one or a few food species (specialists). Inefficient generalists, or consumers that do little to harm their hosts, are not likely to evoke the fourth principle nor the consequent low-frequency cycles.

6.4 SUMMARY

In this chapter we

1. Defined the fourth principle of population dynamics, which recognizes that circular causal pathways can be created by the interaction of populations with elements of their environments.
2. Defined passive, or non-reactive, factors as those that do not change quantitatively in response to the density of the population on which they act.
3. Defined reactive factors as those that change quantitatively (or qualitatively) in response to changes in the density of the population on which they act.
4. Extended the equations describing the second and third principles to predator and prey populations interacting as circular causal systems.
5. Showed how circular causality induces low-frequency cycles in the dynamics of consumer and consumed populations, and that environmental variability can sustain and amplify these cycles.
6. Defined the equilibrium isoclines and dynamics in prey–predator phase space.
7. Showed how the period and stability of predator–prey cycles were affected by the parameters of the two-species model and the dominance of bottom-up or top-down control.

6.5 EXERCISES

1. Simulate the dynamics and plot the isoclines for a reactive predator–prey interaction using equations (6.4) and (6.5) with parameters $A_P = A_N = 1$, $C_P = 0$, $C_N = 0.002$, $C_{PN} = 6$, $C_{NP} = 4$, $F = 0$ in a constant and a variable environment ($s_N = 0.1$). Set the initial prey density near to its carrying capacity and the predator population much lower. Analyse the sensitivity of this model to changes in the value of the predator coefficient of competition $C_{PN} = 1, 0.8$. Discuss the sensitivity of the equations to this parameter. Simulations can be done by hand, on a spreadsheet, or using the *Two-species Modelling and Simulation* program *P2b* in the *PAS* computer software mentioned in the Preface.

2. Simulate the dynamics of predator and prey automata using the rules in Section 6.3.1. This game can be programmed on a computer or you can use the *PAS Game PG3* to simulate predator–prey dynamics on various sized lattices and under different degrees of environmental stochasticity. How does lattice size and variability effect the stability of the space-time dynamics?

3. If you have them, run the *PAS* lessons *PL2* (Predator–Prey Dynamics) and *PL5* (Population Cycles).

7 | The fifth principle: limiting factors

The orbit of any one planet depends on the combined motion of all the planets, not to mention the action of all these on each other. But to consider simultaneously all these causes of motion and to define these motions by exact laws exceeds, unless I am mistaken, the force of the entire human intellect. (I. Newton, 1729)

The first four principles of population dynamics define how populations grow (geometrically), and how the interactions between members of a population (co-operation and competition) and between a population and its environment (circular causality) create the feedbacks that regulate population dynamics. The fifth and final principle recognizes that populations are actually embedded in complex webs of interactions with other biological populations and their physical environments, but that only one or a few of these interactions are likely to dominate the dynamics at any particular time and place. It is a simplifying and organizing principle.

There is no question that natural ecosystems are made up of complex networks of interactions between biotic and physical elements and that this can create a web of first-, second- and higher-order feedbacks (the "curse of dimension" faced by Isaac Newton). Embedded in this ecological web may be a species of interest, say a population of deer, salmon or gypsy moths. The question is, do we need to know all the details of the web before we can understand and predict the dynamics of that particular population, or can we identify one, or a few, interactions that dominate the dynamics?

The same problem confronted Isaac Newton as he worked on his laws of planetary motion. Newton realized that the orbit of each planet is affected by all the other planets in the solar system. Yet he was able to derive accurate laws governing the orbits of planets by considering the single interaction between a planet and its star. In other words, the interaction between star and planet is so dominant that all other interactions could be, for all intents and purposes, ignored. Perhaps Newton's solution also applies to ecological systems? If so, it would certainly make the task of the ecologist and resource manager a lot easier.

7.1 FEEDBACK DOMINANCE

Populations inhabiting complex environments can potentially be affected by many feedback loops created by the web of interactions. However, if all the potential feedback loops simultaneously affected population change, then the dynamics would be complex, unpredictable, and chaotic (this is similar to the "three-body problem" in mathematics where it is virtually impossible to analytically solve three simultaneous, coupled differential equations). However, although the search for chaos in nature has been diligent and exhaustive, the empirical evidence suggests that natural populations fluctuate in a much more regular manner than would be expected if nature were indeed chaotic[21]. This is the first indication that the dynamics of populations of living creatures may be controlled by relatively simple feedback structures.

There are also logical reasons why this should be so. First note that +feedback *and* -feedback cannot dominate the dynamics of a system simultaneously, for their effects counteract each other; i.e. +feedback causes unstable behaviour while -feedback stabilizes. A system cannot be unstable and stable at the same time! Hence, populations must either be dominated by +feedback *or* -feedback at any given time and place. But what if many -feedback loops are present? To answer this question think about a room in which there are two thermostats, each hooked up to separate heaters. In this system, the temperature of the room is regulated by two -feedback loops, each thermostat being engineered to provide -feedback control of temperature. Now visualize how the temperature of a cold room will rise when the thermostats are set at different temperatures. The room will first heat up quickly under the influence of both heaters. However, when the temperature reaches the set point of the lower thermostat, its heater will turn off and temperature change will be driven by the second heater alone. The room will continue to heat up, but more slowly, until it reaches the higher setting. At this point the temperature of the room will be completely controlled by the higher thermostat, and the other one can be removed or destroyed with no effect on the dynamics of the system. On the other hand, should the higher thermostat malfunction, the room will cool to the temperature setting of the lower one. The second thermostat has now taken over control of room temperature. Hence, it seems logical to propose that, out of the complex network of

¯feedback loops acting on a particular population, only one will dominate the dynamics in the neighbourhood of a stabilizing equilibrium, but if that one fails, others will take over the role of population regulation.

7.2 LIMITING FACTORS

The idea that certain feedbacks in the web can dominate the dynamics of natural populations also has roots in experimental ecology. For example, experiments designed to test the effects of fertilizers on agricultural production demonstrated that the nutrient in shortest supply was the one that limited yields. This phenomenon came to be known as the "law of the minimum" or the "law of limiting factors"[25]. Think of it this way: a plant requires two elements for growth, say nitrogen and phosphorus, but one of these, say nitrogen, is present in low concentrations. At first the plant population will grow normally but soon will have problems obtaining enough nitrogen for normal growth. At this point, growth will slow down and eventually stop. In this example, both nitrogen and phosphorus are passive environmental factors for which the population of plants competes according to the third principle. This competition causes the plant R-function to decline with population density to create a stabilizing equilibrium or carrying capacity. In our example, however, we have two feedbacks, one for each resource, which can be represented by two separate R-functions (Figure 7.1). Because nitrogen is more scarce, however, competition for this resource will be more intense, the ¯feedback proportionally stronger, and the slope of the R-function steeper (larger coefficient of *intra*specific competition, C). Hence, the carrying capacity set by nitrogen will be lower than that for phosphorus, and nitrogen will act as the *limiting factor* (Figure 7.1).

It is important to realize that the concept of limiting factors and feedback dominance also applies to enemies and other negative resources. Here we can think of "enemy-free space"[13] as the limiting resource. Hence, predators, parasites and pathogens can also act as limiting factors.

In addition, the concept of limiting factors should not be interpreted too narrowly. In its strict interpretation, the "law of the minimum" only applies to factors that operate independently of each other. However, it is also possible, and perhaps normal, for environmental factors to operate together, as an integrated group, to limit population growth[26]. In these cases the group, not a single species, must be considered as the limiting factor.

Finally, we should be aware that the fifth principle only applies to ¯feedback in the vicinity of a stabilizing equilibrium. The dynamics of populations that are in transition, or that fluctuate widely around their equilibrium points, may be influenced by more than one feedback process.

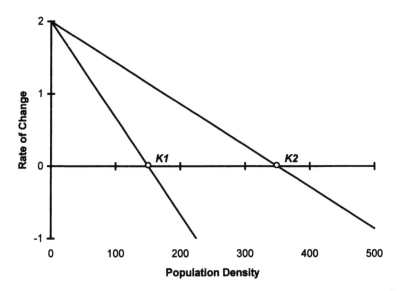

Figure 7.1 Two R-functions resulting from *intra*specific competition for two independent passive resources; the population will be limited at $K1$ by competition for the resource in shortest supply, but if this supply is increased sufficiently so that $K1 > K2$, then the second resource may limit the population at $K2$ [the models are $R1 = 2(1 - N_{t-1}/150)$; $R2 = 2(1 - N_{t-1}/350)$].

7.3 SHIFTING LIMITS

Although the fifth principle postulates that a particular feedback loop will dominate the dynamics of a population at a particular time and place, it does not mean that the dominant feedback cannot change from time to time and place to place. For instance, if a lot of nitrogen is added to the plant system discussed earlier, then the population would resume growing until it became limited by phosphorus or some other nutrient. A new limiting factor has taken over and a new feedback loop (e.g. competition for phosphorus) assumes dominance.

Things become even more complicated if feedback dominance shifts in accordance with the density of the population. For example, when populations are very dense, the hazards of living in a crowd should be greater than the rewards, and the third principle should dominate. On the other hand, when the density is very low, the benefits of living with others should outweigh the hazards, and the second principle should dominate. Under these conditions, the R-function can have a positive slope at low densities and a negative slope at high densities (Figure 7.2). Notice that this R-function can have two equilibrium points, an unstable underpopulation

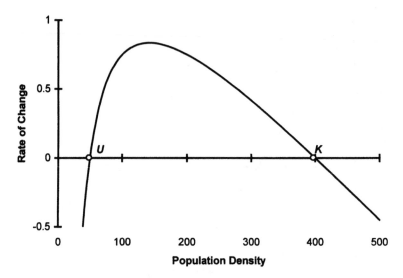

Figure 7.2 Form of an R-function when the second principle dominates at low population densities and the third principle dominates at high densities, where U is an unstable extinction threshold and K is a stabilizing equilibrium or carrying capacity [the model is $R = 2(1 - 150/N_{t-1})(1 - N_{t-1}/400)$].

threshold (U), created by the dominance of the second principle, and a stabilizing high-density equilibrium or carrying capacity (K), created by the dominance of the third principle. Hence, the actual shape of the R-function for a particular population depends on how the two diametrically opposed forces of competition and co-operation change with respect to population density; i.e. a positive slope to the R-function indicates ⁺feedback due to the domination of the second principle, while a negative slope indicates ⁻feedback due to the domination of the third principle.

If general predators with sigmoid behavioural responses[10] are present in the environment, it is possible for a prey population to have an R-function with three (or more) equilibrium points (Figure 7.3):

1. A low-density stabilizing equilibrium (J) created by the switching and/or aggregation behaviour of the predators (see Figure 5.2). Here predators act as the limiting factor with competition for "enemy-free space" the dominant ⁻feedback process. R-functions created by this mechanism will have a negative slope at low densities (remember that several species of predators may operate together as a limiting guild).
2. An unstable escape threshold (E) created by predator satiation, or incidental co-operation amongst prey. Over this density range the R-function has a positive slope because ⁺feedback dominates and limiting factors are overruled.

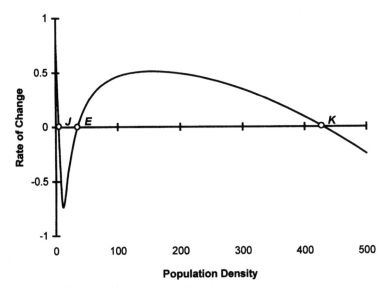

Figure 7.3 A complex *R*-function created by predators that switch and/or aggregate in response to prey density creating ⁻feedback at very low densities and ⁺feedback at intermediate densities. At very high densities, prey numbers are limited by ⁻feedback due to competition for food or other limiting resources. The *R*-function has two stable (*J*, *K*) and one unstable (*E*) equilibrium. The model is $R = 0.8(1 - [N_{t-1}/450]^2) - 30N_{t-1}/(100 + N_{t-1}^2)$.

3. A stabilizing high-density equilibrium or carrying capacity (*K*) created by *intra*specific competition for another limiting resource such as food or space. Here the *R*-function once again has a negative slope because ⁻feedback resumes dominance, but with a different limiting factor.

As we have seen, the dominance of a particular feedback process can shift because of changes in the relative abundance of limiting factors or in the density of the population. In addition, even though one feedback loop may dominate the dynamics at a particular time and place, others may be "waiting in the wings" in case the dominant one fails (remember the thermostats). For example, some insect populations are limited by insectivorous vertebrates when their densities are low, by insect parasitoids if they escape from vertebrate limitation, by pathogens if they escape parasitoid limitation, and by food in the absence of all the above. Feedback *hierarchies* created in this way can provide the ecosystem with a high degree of temporal stability, as well as insurance against the failure of any particular stabilizing mechanisms[26].

7.4 MATHEMATICAL INTERPRETATION

When a population is embedded in a web of interactions with other components of its environment, its R-function can be specified by a general equation such as

$$R = f(N, X_1, X_2, X_3, \ldots X_i; Y_1, Y_2, Y_3, \ldots Y_j), \qquad (7.1)$$

where N is the density of the subject population, $X_1, X_2, \ldots X_i$ are the values of reactive environmental elements, and $Y_1, Y_2, \ldots Y_j$ are the values of passive environmental elements. If only passive factors are present, then the R-function will be dominated, at least in the vicinity of a stabilizing equilibrium, by a first-order feedback process (Chapter 5), which can be specified by the general feedback equation

$$R = f(N_{t-1}). \qquad (7.2)$$

The arbitrary function f can be replaced by the non-linear logistic equation (5.8) or any other reasonable first-order model (some other models will be discussed in Chapter 10). Of course, the problem for the ecological detective is to determine which of all the possible limiting factors was responsible for the first-order feedback. On the other hand, when reactive factors are present in the environment, equation (7.2) is no longer valid because the fourth principle can be evoked. However, it can be shown that the presence of circular causal pathways introduces *time delays* into the feedback loops[24], so that the general R-function for a population competing for passive and reactive resources can be written

$$R = f(N_{t-1}, N_{t-2}, N_{t-3}, \ldots N_{t-d}), \qquad (7.3)$$

with d the maximum delay created by the feedback loop involving the most reactive factors. The maximum time delay parameter d is sometimes called the *dimension* of the system. If we assume that this function has the same basic form as the logistic equation (5.8), then we can write the multiple-delay non-linear logistic equation

$$R = A - C_1 N_{t-1}^{Q_1} - C_2 N_{t-2}^{Q_2} - \cdots C_d N_{t-d}^{Q_d}. \qquad (7.4)$$

Note that populations governed by the first three principles have dimension one ($d = 1$), while the fourth principle allows dimensions greater than unity ($d > 1$).

We can simplify this equation by appealing to the fifth principle. Because only one feedback is likely to dominate the dynamics of a population close to a stabilizing equilibrium, then equation (7.4) can be approximated, under these conditions, by a single-delay non-linear logistic equation

$$R = A - C_{t-d}^{Q}, \qquad (7.5)$$

where d is now the delay created by the dominant feedback, or the dominant dimension.

While on the subject of time delays, it should be noted that delays in the feedback process can be created by mechanisms other than interactions with reactive environmental factors. For instance, if competition amongst parents for passive resources leads to changes in the reproduction and/or survival of their offspring, delays can be introduced into the feedback acting on population density[27]. This is sometimes called the *maternal effect*. In addition, if population density induces *qualitative* changes in resources or enemies, delayed feedback can also be created. For example, some leaf-eating insects induce changes in the quality of their future food supplies when densities are high, and this feeds back to affect future generations.

7.4.1 Complex *R*-functions

On occasion it may be necessary to build models for *R*-functions with unstable thresholds (Figures 7.2 and 7.3). The problem is that natural populations are almost never observed near unstable equilibria, for they quickly move away from these points. Thus, data are rarely available for validating or fitting *R*-functions in regions where the second principle dominates. On the other hand, populations often spend a great deal of time near to stabilizing equilibria, and so data are frequently available for validating or fitting *R*-functions describing ¯feedback due to the third and/or fourth principles. For these reasons, real populations are usually described by *R*-functions with negative slope. When stabilizing processes are known to shift because of the action of the second principle, the new limiting process can be described by a separate *R*-function, and the location of the unstable threshold determined by indirect methods, including guesswork (Figure 7.4). Under these conditions, the model for a system with two stabilizing equilibria can be written

$$R = \begin{cases} A - C_K N_{t-d_K}^{Q_K} ; N_{t-1} > E \ldots \text{high-density} \\ A - C_J N_{t-d_J}^{Q_J} ; N_{t-1} < E \ldots \text{low-density} \end{cases} \tag{7.6}$$

where the subscripts K and J identify the parameters associated with the high-density and low-density *R*-functions, respectively, and E is the escape threshold. Note that the two *R*-functions can have different time delays, enabling the fourth principle to operate around either or both.

7.5 POPULATION DYNAMICS UNDER THE FIFTH PRINCIPLE

The fifth principle recognizes that one feedback process involving a single limiting factor, or co-acting group of factors, will usually dominate the

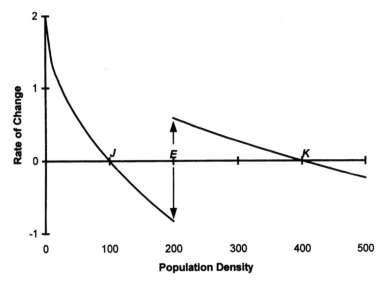

Figure 7.4 Approximation of a complex R-function by a system of two stabilizing R-functions; a high-density R-function with equilibrium K and a low-density R-function with equilibrium J. The decision on which R-function to use depends on whether the current density is above or below the escape threshold E; i.e. when $N_{t-1} > E$, the population obeys the high-density equation and when $N_{t-1} < E$, it obeys the low-density equation. The model is $R1 = 2(1 - [N_{t-1}/400]^{0.5})$ if $N_{t-1} > 200$; $R2 = 2(1 - [N_{t-1}/100]^{0.5})$ if $N_{t-1} < 200$.

dynamics of natural populations in the vicinity of stabilizing equilibria. Under these circumstances, the dynamics can be described by single-delay models, such as the logistic equation (7.5). Simulations with this model show that time delays (or dimension) destabilize the equilibrium and lead to cycles of increasing amplitude and period (Figure 7.5). Notice that, as expected, populations regulated by high-dimensional feedback loops exhibit regular low-frequency cycles similar to systems of mutually interacting populations (Chapter 6).

It is now possible to state the general stability properties of the generalized logistic equation:

1. Smooth approach to equilibrium when $AQd \leq 1$;
2. Damped-stable approach to equilibrium when $1 < AQd < 2$;
3. Periodic cycles around equilibrium when $2 \leq AQd \leq 2.692$;
4. Aperiodic chaos when $AQd > 2.692$.

The dynamics of populations regulated by two stabilizing feedback loops separated by an escape threshold can be simulated with equation (7.6). When both R-functions are stable $(AQd < 2)$ and there are no external environment disturbances, the dynamics will be completely determined by

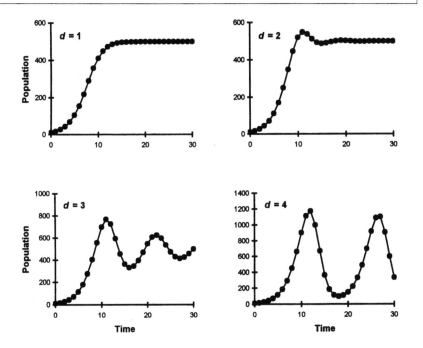

Figure 7.5 Effect of time delays on the dynamics of the logistic equation (7.5) when the parameters are $A = 0.5$, $C = 0.001$, $Q = 1$, and d varies as shown.

the initial density of the population, N_0; i.e. the population will be drawn towards the high-density equilibrium if $N_0 > E$ and towards the low-density equilibrium if $N_0 > E$. In a variable environment, however, random disturbances may displace the population across the escape threshold at unpredictable times resulting in unpredictable switches from low-density dynamics to high-density dynamics (Figure 7.6).

The dynamics can become even more complicated if time delays are present in the R-functions. In this case cyclical instability can be induced around the equilibrium points ($AQd \geq 2$) resulting in more frequent transfers across the escape threshold (Figure 7.6, bottom). Finally, emigration from high-density local populations can raise the density of surrounding populations above their escape thresholds resulting in the spread of high-density dynamics into adjacent areas and, eventually, over huge regions (this phenomenon will be discussed further in Chapters 8 and 14).

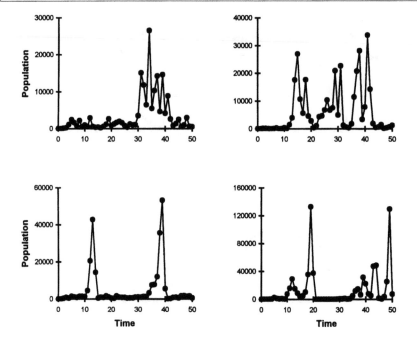

Figure 7.6 Population dynamics predicted by the two equilibrium equations (7.6) when $A = Q_J = Q_K = 1$, $J = 1000$, $K = 10,000$, $E = 3000$ with $d_J = d_K = 1$ (top) and $d_J = 1, d_K = 2$ (bottom) and random variation $s = 0.6$ (left) and $s = 1.0$ (right).

7.6 SUMMARY

In this chapter we

1. Defined the fifth principle of population dynamics which recognizes that populations are embedded in a tangled web of feedback loops, but that only one is likely to dominate the dynamics near equilibrium.
2. Defined the limiting factor as that passive or reactive factor, or group (guild) of co-acting factors, involved in the dominant feedback loop.
3. Showed how limiting factors can change in response to changing environmental conditions and population density.
4. Showed how complex R-functions can be formed if the relative dominance of *intra*specific competition and co-operation change with population density.
5. Described how populations can be regulated in time and space by hierarchies of limiting factors that come into effect at different population densities.

6. Showed how the dimension of the system (number of elements in the web) can be described by time delays in the R-function for a given population.
7. Showed how increasing time delays or dimension destabilizes population dynamics, causing oscillations of increasing amplitude, period, and complexity.
8. Showed how populations governed by complex R-functions with more than one stabilizing equilibrium can exhibit unpredictable population dynamics, including collapses to very low densities and explosions to very high densities, and that these effects can spread over large areas.

7.7 EXERCISES

1. Simulate the dynamics of a population* obeying the time-delayed logistic equation (7.5) when $A = 1$, $C = 0.001$, $Q = 1$ and $d = 1, 2, 3, 4$ in a constant and noisy environment.
2. Simulate the dynamics of a population* governed by two R-functions separated by an unstable threshold [equation (7.6)] when the parameters for the high-density R-function are $A = Q_K = d_K = 1$, $K = 10,000$, for the low-density R-function are $Q_J = d_J = 1$, $J = 1000$, the escape threshold is $E = 3000$, and the initial density of the population is 3000. Run many simulations with different levels of environmental noise.
3. Repeat exercise 2 with all parameters the same except $d_K = 2$.

*These problems can be done on a typical computer spreadsheet or using the *PAS* software program *P1b – Single-species Modelling Simulation* (see Preface for details).

PART TWO

Analysis

Classification: ordering nature's rhythms

If each of the many things in the world were taken as distinct, unique, a thing in itself unrelated to any other thing, perception of the world would disintegrate into complete meaninglessness. (G. G. Simpson, 1961)

In Part One of this book we saw how interactions between members of a population and between populations and their environments determine the expression of the principles of population dynamics and the consequent patterns of population fluctuations. In Part Two we apply this knowledge to the practical problem of understanding and managing real populations. Of the many ways in which theory can be applied to practical problems, one is the development of *classification systems*.

8.1 CLASSIFYING POPULATION DYNAMICS

Classification systems attempt to organize things that we see or experience into groups or classes according to their common characteristics. In the case of population dynamics, the phenomena that we observe and would like to categorize are patterns of population fluctuations. Population dynamics theory can help in this endeavour.

One of the first things we learned in this book was that, at the most basic level, patterns of population fluctuations are the result of *endogenous* and *exogenous* factors acting on the birth and death rates of individual organisms. Therefore, let us make this the first dichotomy in the classification scheme.

8.1.1 Endogenous dynamics

Endogenous patterns of population dynamics result from the action of density-induced *feedback loops* on the birth and death rates of individual organisms. Because there are two basic kinds of feedback, one destabilizing and the other tending to stabilize population dynamics, it seems logical to separate endogenous patterns of population dynamics into those generated by +feedback, which we will call *unstable dynamics*, and those generated by ‾feedback, which we will call *stable dynamics*. The use of the term "stable dynamics" should not be interpreted to mean that there is a stable equilibrium point, but rather that the population tends to return towards an equilibrium point or average density (remember from Chapter 5 that ‾feedback is a necessary but not sufficient condition for stability of the equilibrium point).

Unstable patterns of population dynamics result from the action of the first and second principles. When the first principle acts alone, populations grow or decline exponentially (Figure 3.1), while if the second principle is also involved, they grow or decline at a faster than exponential rate, what has been called hyper-exponential growth (Figure 4.3). In addition, the second principle often creates an unstable *extinction threshold* separating hyper-exponential growth from hyper-exponential collapse to extinction. Thus, unstable population patterns can be divided into two classes, *exponential* and *hyper-exponential dynamics*, depending on whether the second principle is involved or not.

Stable patterns, on the other hand, result from the action of the third and fourth principles of population dynamics. Once again there are two basic kinds of dynamics, the high-frequency or saw-toothed oscillations resulting from the dominance of the third principle alone (first-order dynamics as illustrated by Figure 5.5), and the low-frequency or "cyclical" oscillations observed when the fourth principle is also operating (higher-order dynamics as illustrated by Figure 6.4). Thus, stable dynamics can also be divided into two classes identified as *first-order* and *higher-order* dynamics.

Finally, there are populations in which +feedback and ‾feedback can switch dominance as population density changes, giving rise to complicated dynamic patterns (Figure 7.6). Populations with these characteristics can be in the stable class at certain times and the unstable one at others, so that we do not really need to define another class. However, this kind of population dynamics is so unique and important that it probably deserves a class of its own called *meta-stable dynamics*[28]. There seem to be two plausible patterns of meta-stable behaviour: in the first, the third principle dominates at low and high densities so that we have two stabilizing equilibria and the dynamics are characterized by sustained periods at high density (Figure 7.6, top). In the second, the fourth principle dominates at high densities so that a single high-density cycle is usually observed (Figure 7.6, bottom). These two behaviours will be called *bi-stable* and *pulse* dynamics, respectively. Another theoretically possible pattern involves the fourth principle operating at low densities. However, because there are no examples of this kind of pattern in nature, it will be ignored in the classification system (if fact it may not be biologically plausible for the fourth principle to operate at low population densities in meta-stable systems).

8.1.2 Exogenous dynamics

Exogenous dynamics can result from random or directed environmental factors or inputs. Random inputs are those that vary with time but the mean value of the input does not change in any kind of systematic manner. As a result, populations tend to exhibit *random variation* around an average trajectory (Figure 5.7). On the other hand, directed environmental inputs cause population parameters to change in a consistent (non-random) manner so that the mean levels of abundance and degree of fluctuation can change with time. This will be called *forced dynamics*. There are two basic kinds of forced dynamics; those called *trends*, where the external input is applied gradually (Figure 5.11), and those called *steps*, where the input changes suddenly (Figure 5.12).

We have now identified and classified all the patterns of population dynamics predicted by the principles of population dynamics. The classification scheme is summarized in Table 8.1. It should be remembered, however, that classifications are human attempts to order nature and, therefore, that there may be better ways to do this.

8.2 CLASSIFYING PEST OUTBREAKS

Outbreaks of pests have devastated crops and forests, and threatened human health and well-being, from time immemorial. Obviously, it would be helpful to recognize different classes of pest outbreaks, for then it might be possible to choose the correct response for the particular class of outbreak threatening us or our resources. From the classification of population dynamics derived above (Table 8.1), it is apparent that there are two ways in which a population of pests can attain outbreak densities: the first is through a change in the exogenous environment that raises the equilibrium density of a species to such an extent that it causes significant damage. In this situation, the dynamics of the pest population are *forced* by external factors so that the average pest density is a graded response to external factors. These will be called *gradient outbreaks*[29]. One of the distinguishing features of gradient outbreaks is that they are

Table 8.1 A classification for population dynamics

1. **Endogenous dynamics** – due to feedback processes
   1A. **Unstable dynamics** – due to ⁺feedback
      1Aa. *Exponential dynamics* – due to the operation of the first principle alone
      1Ab. *Hyper-exponential dynamics* – due to the combined action of the first and second principles; extinction thresholds are possible
   1B. **Stable dynamics** – due to ⁻feedback
      1Ba. *First-order dynamics* – due to the third principle; saw-toothed oscillations
      1Bb. *Higher-order dynamics* – due to the fourth principle; cyclical oscillations
   1C. **Meta-stable dynamics** – due to a combination of ⁺feedback and ⁻feedback
      1Ca. *Bi-stable dynamics* – due to the third principle operating at high densities
      1Cb. *Pulse dynamics* – due to the fourth principle operating at high densities
2. **Exogenous dynamics** – due to external (non-feedback) processes or inputs
   2A. **Random variation** – due to unpredictable changes in the exogenous environment
   2B. **Forced dynamics** – due to predictable changes in the exogenous environment
      2Ba. *Trend* – gradual change in the average density or pattern of population dynamics
      2Bb. *Shift* – sudden change in the average density or pattern of dynamics

restricted to certain locations where the environment is particularly favourable for the pest species. These outbreaks do not spread into adjacent unfavourable habitats because the population automatically adjusts to the equilibrium density determined by the limiting factor in the new habitat.

In contrast, if pest populations exhibit meta-stable dynamics, the outbreak can spread out from local *epicentres* to cover huge areas if dispersal from the epicentre is sufficient to raise nearby populations above their escape thresholds (e.g. see Figure 14.2). Because of this expansive and invasive tendency, outbreaks of pests with meta-stable dynamics have been termed *eruptive*. Thus, the first and most important division of outbreaks is into gradient and eruptive types.

Gradient outbreaks are usually observed in certain favourable habitats, or susceptible sites, where populations may remain persistently high if the dynamics are first order or go through regular cycles of abundance if the dynamics are of higher order. These will be called *sustained gradients* and *cyclical gradients*, respectively. On the other hand, environments may sometimes change from unfavourable to favourable and back again, in which case the pest population will follow a "boom and bust" trajectory. These will be called *pulse gradients*.

Eruptive outbreaks can also persist for some time (*sustained eruption*), particularly if the dynamics around the high-density equilibrium are first order (Figure 7.6, top) or they can also go through a rapid cycle of destruction and collapse (*pulse eruption*) if the outbreak dynamics are higher order (Figure 7.6, bottom). It is also theoretically possible for higher-order dynamics to occur around the lower equilibrium, in which case we may observe *permanent* or *cyclical eruptions*. As noted earlier, however, there do not seem to be any examples of this kind of outbreak and, as permanent outbreaks would destroy the environment, they do not seem biologically plausible. The outbreak classification scheme is summarized in Table 8.2.

Table 8.2 A classification of pest outbreaks[29]

1. Outbreaks do not spread out from local epicentres	**Gradient** outbreak
R-function has single stabilizing equilibrium	
Second principle is not evoked at moderately high densities	
1A. High-frequency or saw-toothed oscillations	
First-order dynamics	
Third principle dominates	
1Aa. Outbreaks permanently associated with particular localities	**Sustained** gradient
1Ab. Outbreaks follow temporary environmental disturbances	**Pulse** gradient
1B. Low-frequency or cyclical oscillations	
Higher-order dynamics	
Fourth principle in operation	
1Ba. Outbreaks permanently associated with particular localities	**Cyclical** gradient
1Bb. Outbreaks follow changing environmental conditions	**Pulse** gradient
2. Outbreaks spread out from local epicentres	**Eruptive** outbreak
R-function has two or more stabilizing equilibria	
Second principle is evoked at moderately high densities	
2A. High-frequency or saw-toothed oscillations at sparse densities	
First-order dynamics at lower equilibrium	
Fourth principle not operating at low density	
2Aa. High-frequency oscillations at high density	**Sustained** eruption
2Ab. Cyclical oscillations at high density	**Pulse** eruption
2B. Low-frequency or cyclical oscillations at sparse density	
Higher-order dynamics at lower equilibrium	
Fourth principle in operation at low density	
NOTE: This type of outbreak is theoretically possible but may not be biologically feasible	
2Ba. High-frequency oscillations at high density	**Permanent** eruption
2Bb. Cyclical oscillations at high density	**Cyclical** eruption

8.2.1 Management implications

The ability to recognize different classes of outbreaks has important implications for pest managers, for different kinds of outbreaks require
different management approaches. For example, the use of pesticides to
suppress sustained gradient outbreaks is unlikely to be cost-effective
because the population will quickly grow back to its original density,
requiring repeated pesticide applications. On the other hand, eruptive
outbreaks can be prevented from spreading by vigorous pest control at
the outbreak *epicentre* (the place where the outbreak begins). Under these
conditions it may be cost-effective to use pesticides against the pest
because only one application may be required at the epicentre to prevent
the outbreak from spreading. However, it is imperative that the impending
outbreak is attacked with great vigour and determination, for once
eruptive outbreaks have spread over large areas they are almost impossible to stop.

   In the long run, it is probably better to try and manage outbreak species
by controlling the environment; for example, by decreasing the abundance
of food or shelter for the pest species, increasing the abundance of natural
enemies, or by only growing crops in physical environments unfavourable
to the pest. Through activities such as these, the manager may be able to
reduce the equilibrium densities of gradient species to tolerable levels or
deepen the "predator" (or "host resistance") pit (Figure 5.2) and so reduce
the likelihood of eruptive outbreaks.

8.3 CLASSIFICATION BASED ON SPECIES CHARACTERISTICS

Biologists have had an abiding interest in attempting to classify observed
phenomena, such as pest outbreaks, according to the genetic characteristics of the species[30]. The utility of such an approach is obvious, for if there
are certain genetic traits associated with outbreak species, it would be
possible to predict the potential of the species to exhibit outbreaks without
observing population dynamics at all! For example, knowing that the
second principle must operate at relatively high population densities in
eruptive species, and that this principle is evoked by co-operative
behaviour, it is tempting to predict that all outbreak species will have
co-operative attack or defence behaviours, or some kind of aggregation
behaviour. Unfortunately, although many species that exhibit outbreaks
have adapted aggregation behaviours, many others do not. This is because
the second principle can be evoked by *incidental* co-operation which does
not entail any adapted co-operative trait. In addition, there are many
examples of species which possess adapted co-operative and/or aggregation behaviours but which do not exhibit outbreaks. Despite this, there

are certain traits that are frequently enough associated with outbreak species that they bear remembering (Table 8.3). Because much of the work on outbreaking species has been done by entomologists working on forest pests, many of the traits listed in the table should be restricted to insects and, in particular, to Lepidoptera (mainly moths), living in forest environments.

Table 8.3 Characteristics of outbreak species (mostly insects)[30]

1. Species with high reproductive potential
2. Species with high mobility and dispersal abilities (but see below)
3. Species that utilize rare or ephemeral habitats or food supplies
4. Species with well-developed co-operative or aggregation behaviour
5. Insects that are vulnerable to predation by vertebrates that switch/aggregate in response to prey density
6. Insects that rely on food stored by immature stages, rather than adult feeding, for reproduction
7. Insects which do not carefully select the food supply for their offspring
8. Insects that go through colour or phase polymorphism in response to density
9. Insects that hibernate or overwinter in the egg stage
10. Insects with poor dispersal capabilities
11. Insects with broad food preferences

By now it should be evident that the kind of dynamic behaviour exhibited by a particular species depends as much on the characteristics of the other organisms with which it interacts as on its own adaptive traits. Hence, the type of dynamics expected from a given species is difficult to predict from its own genetic characteristics. On the other hand, if one has an intimate knowledge of a species and how it interacts with elements of its environment, it may be possible to deduce the kind of dynamic behaviour it is likely to exhibit. This is the basic idea behind *qualitative modelling* discussed in Chapter 10.

8.4 SUMMARY

In this chapter we

1. Developed a scheme for classifying population dynamics.
2. Developed a scheme for classifying pest outbreaks.
3. Showed how the identification of outbreak types is essential for developing effective pest management strategies.
4. Summarized some of the special characteristics associated with outbreak species.

8.5 EXERCISES

Below are listed some forest insect pests and their outbreak characteristics. Identify the outbreak class from Table 8.2 and suggest a management strategy to avoid or control outbreaks.

1. The larch budmoth feeds on the needles of larch trees through much of Europe without causing significant defoliation. However, in some high-elevation larch forests, such as those of the high Alps, spectacular outbreaks occur, almost like clockwork, every 10 years or so (Figure 1.1).
2. The fir engraver beetle lives in the interior fir forests of western North America, killing fir trees weakened by environmental stresses. Outbreaks occur when whole stands or forests are weakened by severe droughts, outbreaks of leaf-eating insects, or root diseases, but quickly subside when the stress is removed. Outbreaks do not spread into healthy forests.
3. The white pine weevil attacks the terminal shoots of Sitka spruce in the Pacific Northwest of America causing loss of growth and stunted trees. Economically significant damage occurs year after year on sites that are deficient in soil moisture but is inconsequential on wet sites.
4. The western spruce budworm feeds on the needles of fir trees in the interior Pacific Northwest. Outbreaks occur at irregular intervals and often spread over large areas. Outbreaks may persist for several or many years in a particular locality. Birds are thought to be a major factor limiting budworm populations at sparse densities during the long non-outbreak periods.
5. The Douglas-fir tussock moth also feeds on the needles of fir trees in the interior Pacific Northwest. Outbreaks occur at fairly regular 10-year intervals in certain locations but have never been observed in other places, even though fir trees are abundant.
6. The spruce bark beetle infests and kills spruce trees in Scandinavia and other European countries. For much of the time the beetle is restricted to spruce trees that have been felled by loggers and windstorms, or have been severely damaged by root diseases or other environmental factors. Following severe windstorms or prolonged droughts, spruce beetle populations may build up and spread through healthy forests, destroying millions of trees.

Diagnosis: explaining nature's rhythms   9

Two things are crucial to the study of animal population dynamics: to carry out an uninterrupted long-term observation of a natural population and to apply theoretical knowledge to the analysis, interpretation and synthesis of the results. (T. Royama, 1992)

In the last chapter we used our knowledge of population theory to develop a general scheme for classifying population dynamics and pest outbreaks, and explained how the correct identification of the type of dynamics was essential for effective population management. The next task is to develop a methodology for recognizing the kind of dynamics exhibited by real populations, and for detecting the underlying factors responsible for those dynamics, an activity called *diagnosis*.

Diagnosis is the *art* of detecting the cause of a particular condition from observations of its symptoms as, for instance, in the diagnosis of diseases in medicine. In this case, the physician attempts to determine the cause of a disease from clues obtained by measuring certain dynamic variables, such as the patient's blood pressure, blood chemistry, brain wave patterns, and so on. Likewise, resource and pest managers attempt to determine the kind of dynamic behaviour being exhibited by the population they are trying to manage and, if possible, the causes of those dynamics. This process can be facilitated by examining certain statistical clues, or diagnostics, some of which are discussed in this chapter.

Diagnosis is an art because its success rests to a large extent on the skill and knowledge of the practitioner. Even though the diagnostician may use sophisticated techniques, the final diagnosis is no more than an educated opinion. This is why it is often advisable to seek a second opinion on a serious medical problem, and why the same principle should also apply in ecology. Our faith in the diagnosis rests, in the end, on the credibility of the diagnostician – his or her knowledge of facts and theory and absence of prejudice or bias – for it is difficult to trust the opinion of someone who has a vested interest in the results of the diagnosis. To be effective, diagnosticians must have strong ethical standards and must retain their objectivity and credibility at all cost.

Diagnosis is a sensible approach when one is faced with a problem involving a very complicated system, where it is impossible to measure or understand all the quantitative details. This is why diagnosis is employed by physicians and automobile mechanics, and why it is also relevant in applied ecology, for there is no way that we can ever know all the details about a particular ecological system.

9.1 FUNDAMENTALS OF POPULATION DIAGNOSIS

The main objective of an ecological diagnosis is to identify the kind of ecological phenomenon we are observing and, if possible, to understand the underlying cause or causes of that phenomenon. In the case of population dynamics, the first job is to identify the kind or class of dynamics being observed and then, if possible, determine the cause or causes of the observed dynamics. Success in this venture depends on the kind of information available – the *data*.

9.1.1 Population data

In order to make a diagnosis, observations (data) are required on the system being studied. In the case of population systems, relevant data include birth and death rates, numbers of predators present, incidence of disease, and so on. Unfortunately, this kind of data is not usually available to the resource or pest manager. However, as management agencies often keep records of the number of organisms harvested each year, or carry out routine surveys to determine the abundance of certain species on their management units, a series of observations is often available on the number or density of organisms through time. This is called a *time series*. The diagnostic methods developed in this chapter are restricted, in the main, to time series data in one form or another.

Time series data consist of a series of population density estimates taken at equal intervals, usually annually, over a fairly long period of time. Hence, the primary variable available for diagnostic purposes is the number or density of the population at time t, or N_t. However, given estimates of population density at two consecutive points in time, then it is also possible to estimate another variable, the realized per capita rate of change, R, over that period of time. In other words, given that

$$N_t = N_{t-1}e^R,$$

then, after transforming to logarithms, we have

$$\ln N_t = \ln N_{t-1} + R,$$

and rearranging,

$$R = \ln N_t - \ln N_{t-1} = \ln(N_t/N_{t-1}). \tag{9.1}$$

Remember that R can also be estimated from the relationship $R = \ln(1 + B - D)$. Thus, quantitative estimates of two variables, N and R, can be obtained from a simple series of counts commonly made by population managers. In addition, if several estimates of R and N are available, then it may be possible to fit a curve to the data and thereby estimate the parameters of the R-function.

9.1.2 Time series analysis

The basic procedure for time series analysis of population data can be illustrated using data from a laboratory experiment where the effects of exogenous factors have been reduced to a minimum. For example, Table 9.1 shows the number of southern cowpea weevils reared on a constant quantity of azuki beans supplied at the end of each weevil generation (a generation is about 25 days). Notice how the time series fluctuates in the saw-toothed pattern typical of a first-order dynamic process (Figure 9.1, top-left). If the realized per capita rate of change is then calculated (see Table 9.1), and the vectors of population change plotted in N, R phase space, the typical narrow phase portrait of a first-order dynamic process is also obtained (Figure 9.1, top-right; c.f. Figure 5.10).

It is now possible to fit the cowpea weevil data to the logistic R-function [equation (7.5)] using non-linear regression techniques[32]. Notice that the

Table 9.1 Numbers of adult cowpea weevils (N_t) counted at the end of each generation (t; a generation is 25 days) when supplied a constant weight of azuki beans[31] (10 grams every 25 days), and the calculation of the realized per capita rate of change (R)

t	N_t	$\ln N_t$	$R = \ln N_t - \ln N_{t-1}$
1	16	2.773	
2	294	5.683	2.911
3	125	4.828	–0.855
4	250	5.521	0.693
5	130	4.867	–0.654
6	213	5.362	0.494
7	160	5.075	–0.286
8	200	5.298	0.223
9	150	5.011	–0.288
10	180	5.193	0.182

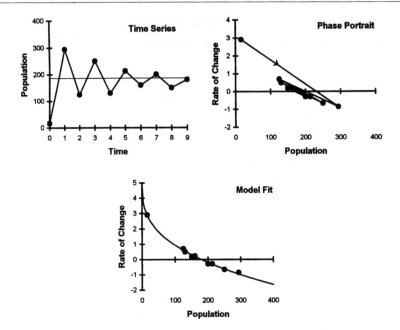

Figure 9.1 Population dynamics of southern cowpea weevils[31] living in a constant supply of azuki beans in the laboratory with the data fitted[32] to a non-linear logistic R-function, $R = A - CN_{t-d}^Q$, with parameters $A = 5.02$, $C = 0.8018$, $d = 1$, $Q = 0.354$ and a coefficient of determination $r^2 = 0.995$.

fitted R-function has a concave form (Figure 9.1, bottom) and that it explains almost all of the variability in the data; i.e. the *coefficient of determination*[32] is $r^2 = 99\%$. It is useful to think about the coefficient of determination as a measure of the contribution of endogenous feedback to the observed population fluctuations. In this sense, the residual variation, $1 - r^2$, can be viewed as the contribution of exogenous random disturbances. Hence, the coefficient of determination enables us to identify the relative importance of endogenous and exogenous factors in determining the observed dynamics.

Unfortunately the analysis of field data is rarely so straightforward. The weevils in our example were provided with constant food and climate while, in the field, weather and food supplies can vary continuously. However, if conditions in the field vary in a non-systematic manner (randomly), then it may still be possible to analyse the data in much the same way, although such large coefficients of determination are unlikely to be obtained, and more sophisticated techniques may be required to recognize the type of dynamic pattern being observed.

Of course, conditions in the field can also vary in a systematic (non-random) way. For example, the parameters can vary smoothly with time, creating trends, or suddenly, creating discontinuities in the time series (remember Figures 5.11 and 5.12). However, the successful analysis of endogenous dynamics, as we have just done for the pea weevil, demands that the time series be *stationary*. Stationarity implies that the mean of the time series does not vary in any systematic manner, although it may vary randomly. Thus, before the endogenous dynamics can be analysed, we must first test the time series to see if it is stationary and, if it is not, the data must be manipulated in some way to make it approximately stationary. It is also important to realize that meta-stable systems can also give rise to non-stationary time series, and that such data also require special analytical procedures, as described later in this chapter.

Finally, the search for endogenous causes of population dynamics relies on the intelligent application of time series analysis. It makes no sense to look for endogenous causes in data that are obviously driven by exogenous factors. For example, seasonal climatic patterns often give rise to regular cycles of abundance as organisms reproduce in summer and die off in winter. This is why annual counts, which ignore seasonal variation, are preferable for time series analysis. As a general rule, time series analysis of endogenous dynamics should only be carried out after a search for exogenous causes has proven unproductive.

9.2 DIAGNOSTIC TOOLS

In this section we discuss some of the diagnostic tools that can help determine if a time series is stationary and, given this condition, to detect the order of the endogenous dynamics.

9.2.1 The time series

Naturally, the first thing to look at is the raw time series because trends, discontinuities, and the type of oscillation are often apparent from a cursory examination of the population trajectory through time. For instance, it is fairly obvious from the pattern of oscillations in the cowpea weevil time series that the mean is relatively constant and that the sawtooth fluctuations are probably the result of a first-order dynamic process (Figure 9.1).

Let us now revisit some time series from natural populations first seen in Chapter 1 (Figure 9.2). Notice that the human population is non-stationary, and that it exhibits a roughly linear logarithmic increase for the first 100 years or so. In the 1800s, however, the population began to grow at an accelerating logarithmic rate. On the other hand, the blue whale series, which is also non-stationary, declined exponentially at first but then seemed to reach a plateau before declining rapidly in the last

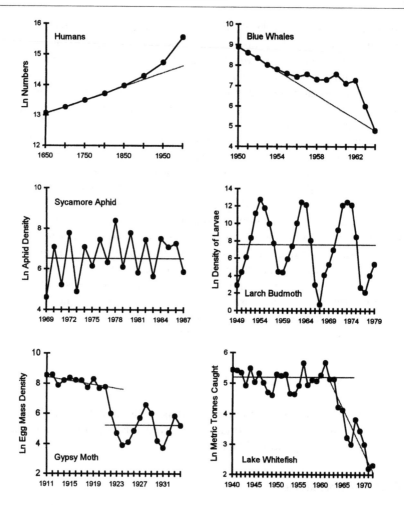

Figure 9.2 Time series data for several natural populations[3] with numbers or densities plotted as natural logarithms and lines showing the expected geometric progression (top graphs) or the mean of the series, including possible trends in the mean.

2 years. The sycamore aphid series appears to be stationary and to have the typical saw-tooth fluctuations of a first-order dynamic system, similar to the cowpea weevil, while the regular cyclic fluctuations of the larch budmoth indicate the presence of higher-order dynamics. Finally we have two non-stationary time series that exhibit sudden changes in behaviour. In the case of the gypsy moth, the population fluctuated at very high densities for 10 years following its introduction into North America, but suddenly declined to a much lower average density. In addition, the oscillations seemed to change from a predominantly first-order pattern to a

higher-order pattern at about the same time. In the last example, the whitefish population of Lake Ontario was relatively stationary for many years and exhibited symptoms of apparent first-order dynamics, but suddenly began to decline over the final 10 years.

As we have seen, a preliminary examination of the time series can provide provisional information about stationarity and the kind of population fluctuations. The other diagnostic tools discussed below may strengthen, weaken or radically alter this original diagnosis.

9.2.2 The return time

If a dynamic variable is displaced from a stabilizing equilibrium by a random shock, it will usually move back towards that equilibrium point after a certain period of time (provided, of course, that it has not been displaced beyond an unstable threshold). The time it takes to return to equilibrium following a disturbance is called the *return time*.

Assume that the *mean*[32] of a stationary time series is a reasonable estimate of the equilibrium point, then the return time can be estimated by counting the number of time steps it takes for the trajectory to reach the mean following a displacement to either side; i.e.

$$RT_m = CTS_m + FTS_m, \tag{9.2}$$

where RT_m is the return time of the mth segment of the time series, CTS_m is the number of complete time steps taken before reaching the mean, and FTS_m is the fraction of time it takes to reach the mean in the final step

$$FTS_m = \frac{N_p - \bar{N}}{N_p - N_a},$$

where $\bar{N}$ is the mean of the series, N_p is the density of the population prior to crossing $\bar{N}$, and N_a is the density after crossing $\bar{N}$.

For example, the mean of the stationary pea weevil series is $\bar{N} = 171.8$. Starting with the first data point ($N_0 = 16$), which is below the mean, it is easy to see that the next data point ($N_1 = 294$) is above the mean. In other words, the trajectory crosses the mean in less than one time step. Hence, $CTS_1 = 0$ and

$$FTS_1 = (16 - 171.8)/(16 - 294) = 0.56.$$

Therefore, the first return time is

$$RT_1 = CTS_1 + FTS_1 = 0 + 0.56 = 0.56,$$

units of time. The second segment begins with $N_2 = 294$ and crosses again before the next datum $N_3 = 125$. Hence, $CTS_2 = 0$, $FTS_2 = (294 - 171.8)/(294 - 125) = 0.72$, and $RT_2 = 0.72$. After calculating the return times for each segment [there are nine segments in the pea weevil series (Figure 9.1)], the mean return time, MRT, is calculated by adding all the individual return times and dividing by the number of segments, M,

$$MRT = \frac{\sum_{m=1}^{M} RT_m}{M}.$$ (9.3)

The variance[32] of the return time is then

$$VRT = \frac{\sum_{m=1}^{M} (RT_m - MRT)^2}{M - 1}$$ (9.4)

In the case of the pea weevil $MRT = 0.575$ and $VRT = 0.027$. Short return times, of course, are typical of rapid, first-order dynamics. Notice that the variance in the pea weevil return time is much smaller than the mean, indicating that there is very little variation in the period of fluctuation over the entire time series.

As another example, consider the larch budmoth time series (Figure 9.2). The logarithms of the first four data points are 2.89, 4.407, 6.10, 8.337 and the mean of the logarithmic series is 7.359. Thus, it takes two complete time steps ($CTS_1 = 2$) plus the fraction [$FTS_1 = (6.10 - 7.359)/(6.10 - 8.337) = 0.563$] to reach the mean, so that $RT_1 = 2.563$. The return time statistics for the budmoth time series are $MRT = 3.856$, $VRT = 0.862$. Once again, the variance of the return time is much smaller than the mean, indicating a relatively constant period of oscillation.

The mean and variance of the return time provide useful clues to the order, or time delay, of the dominant feedback process acting on the dynamics of a population, as well as an indication of the stationarity of the mean. In general

1. $MRT < 2$ implies first-order dynamics ($d = 1$);
2. $MRT > 2$ suggests higher-order dynamics ($d > 1$);
3. $VRT \ll MRT$ indicates a relatively constant period of oscillation; and
4. $VRT \gg MRT$ implies that the time series is probably non-stationary.

Return time statistics for all populations studied in this chapter are given in Table 9.2. Notice that the blue whale, gypsy moth and whitefish time series are all diagnosed as non-stationary because of high variances in their return times.

9.2.3 The autocorrelation function

When a dynamic variable is involved in a feedback loop, the value of that variable at one point in time is affected to some degree by its value at

Table 9.2 Diagnostics for the time series discussed in the text

Population	MRT	VRT	ACF	PRCF
Cowpea weevil	0.575	0.270	Stationary	lag 1
Human	*	*	*	*
Blue whale	1.661	6.049	Non-stationary	*
Sycamore aphid	0.611	0.270	Stationary	lag 1
Larch budmoth	3.856	0.862	Stationary	lags 1 & 2 (3?)
Gypsy moth	5.603	27.770	Non-stationary	*
Sequence no. 1	1.313	1.308	Non-stationary	*
Detrended	1.196	0.556	Stationary	lag 1
Sequence no. 2	2.196	1.106	Stationary	lags 1 & 2 (2?)
Lake whitefish	6.759	39.744	Non-stationary	*
Sequence no. 1	1.852	0.750	Stationary	lag 1
Sequence no. 2	1.835	2.780	Non-stationary	*
Detrended	1.706	1.508	Stationary	lag 1

*Cannot be calculated because the time series is not stationary. ? Best single lag regression.

previous points in time, with the time delay between cause and effect being dependent on the length (order) of the feedback loop [see, e.g. equation (7.3)]. Under these conditions the variable will be correlated with itself. This is called *autocorrelation*. The autocorrelation in a time series is obtained by calculating the *correlation coefficient*[32] between counts separated by a given time delay or time *lag*. For example, the autocorrelation between counts separated by one time step (lag 1) is obtained by calculating the correlation coefficient between N_t and N_{t-1}, or more usually the correlation between the natural logarithms of these quantities (i.e. the correlation between $\ln N_t$ and $\ln N_{t-1}$). This is called autocorrelation at lag 1. In a similar way, autocorrelation at lag 2 is obtained by calculating the correlation between $\ln N_t$ and $\ln N_{t-2}$, and so on. A histogram showing the auto-correlations at lags 1, 2, 3, . . . is called the *autocorrelation function* or ACF (Figure 9.3). Notice how the ACFs for the pea weevil and sycamore aphid switch from a negative to a positive correlation at successive lags, and that the value of the correlation coefficient, or the amplitude of the ACF, decays with increasing lag. ACFs like these are typical of stationary time series generated by rapid, first-order, ¯feedback ($d = 1$) and are sometimes called *phase forgetting*. An ACF that does not decay with increasing lag (*phase remembering*) indicates that the oscillations were driven by an oscillating exogenous variable. Notice that the ACFs for stationary time series oscillate in a systematic and *balanced* way around zero correlation.

Stationary dynamics governed by ¯feedback processes that operate with a longer time delay ($d > 1$) produce more complicated ACFs, but they still fluctuate in a systematic and balanced manner around zero correlation, and their amplitudes generally decay with increasing lag (see the larch budmoth ACF in Figure 9.3). One useful quality of the ACF is that the period of oscillations in the time series is reflected by the period of

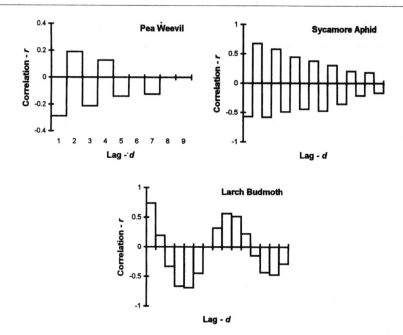

Figure 9.3   Balanced autocorrelation functions (ACFs) for stationary time series.

the ACF; for example, the ACFs for the weevil and aphid time series repeat themselves every two lags (period 2), while that for the budmoth repeats itself every ninth lag (period 9).

ACFs from non-stationary time series tend to decrease continuously with increasing lag rather than fluctuating around zero correlation (Figure 9.4). Other non-stationary ACFs may fluctuate around zero but tend to be *unbalanced* in comparison with ACFs for stationary series.

In general, the ACF provides clues to

1. whether the series is *stationary* or not (balanced versus unbalanced ACF);
2. the period of oscillation (period of ACF);
3. whether the oscillations are driven by endogenous (ACF decays with increasing lag) or exogenous (ACF does not decay with increasing lag) factors and, if the former
4. whether the dominant feedback response is rapid (ACF period 2) or delayed (ACF period > 2).

9.2.4 The phase portrait

Information on the kind of dynamic pattern exhibited by a time series can also be obtained by plotting the data in *N-R* phase space to produce a *phase portrait*. This involves plotting each value of *R* on the initial population

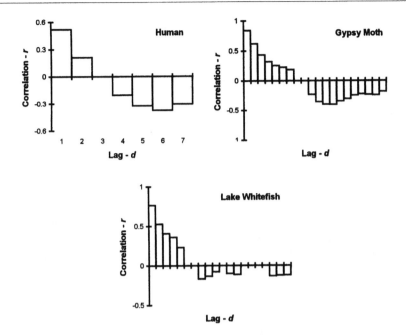

Figure 9.4 Unbalanced autocorrelation functions (ACFs) for non-stationary time series.

density, N_{t-1} or $\ln N_{t-1}$, and then drawing a line (the vector) between successive points. For example, the first coordinate (N, R) for the cowpea weevil is $(16, 2.911)$ and the next is $(294, -0.855)$ (Table 9.1). After plotting these two points in N-R phase space, the vector is drawn between them (Figure 9.1). An arrow can be drawn on the vector to show the direction of change. Notice that all vectors in the pea weevil series point towards, or pass close to, the same place on the $R = 0$ line. This is the typical *narrow* phase portrait of a first-order dynamic system, with all vectors pointing towards, or passing close to, the equilibrium point (see Figure 5.10).

Figure 9.5 shows the phase portraits for four of the time series in Figure 9.2. Notice that, as expected, the sycamore aphid has the narrow phase portrait of a first-order dynamic system, while the larch budmoth has a *broad* circular clockwise phase portrait typical of higher-order dynamics (see Figure 6.2). Notice that the phase portrait for the gypsy moth shows signs of first-order dynamics at high densities and higher-order dynamics at low densities. Finally, the whitefish phase portrait shows evidence of first-order dynamics followed by an eratic collapse.

When dealing with time series with circular phase portraits, information on the order of the dominant feedback process can sometimes be obtained by plotting R on N_{t-d} or $\ln N_{t-d}$ where d is increased until the circular orbit vanishes, or is least evident. For instance, a time delay of $d = 3$ seems

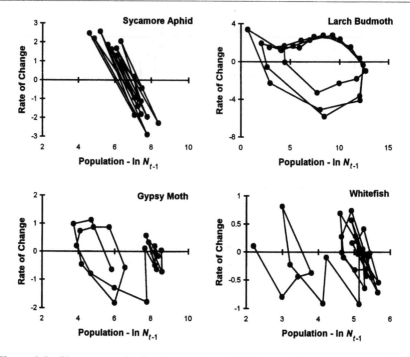

Figure 9.5 Phase portraits for time series exhibiting equilibrium dynamics.

to compress the larch budmoth data the best (Figure 9.6). However, it may be difficult to decide between a delay of 2 or 3 in this case. It is important to notice that the plot of R on $\ln N_{t-4}$ causes the orbit to widen again and to circle in an *anticlockwise* direction. This is a clear indication that a delay of $d = 4$ is too long.

Phase portraits can also be used to detect discontinuities in the time series. For example, the gypsy moth data clearly show a sudden shift from high-density to low-density dynamics (Figure 9.5). In addition, it is quite obvious that the order of the dynamics changed during this shift in the mean density, indicating that a different kind of ⁻feedback mechanism was involved in regulating the gypsy moth population.

In general, the phase portrait can provide useful information on the order of the dominant ⁻feedback process, and whether shifts occurred in the equilibrium density.

9.2.5 The partial rate correlation function

Plotting the phase portrait in time delay coordinates (Figure 9.6) is a useful technique for detecting the dominant feedback delay in a time series provided that the trajectory does not wander too far from equilibrium.

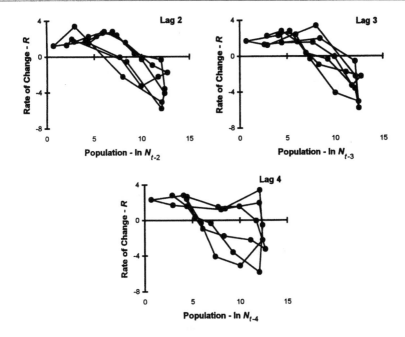

Figure 9.6 Phase portraits for the larch budmoth when R is plotted on $\ln N_{t-d}$ with the time lag d varying from 2 to 4.

Under these conditions the dynamics should be dominated by a single feedback mechanism (i.e. the fifth principle should hold). In the case of the larch budmoth, however, the population fluctuates through several orders of magnitude (from 0.002 to 294 larvae per kilogram of larch foliage) and it becomes possible for more than one feedback to be involved in the dynamics. The *partial rate correlation* can help us to detect such multiple feedbacks.

When the per capita rate of change, R, for the larch budmoth was plotted on lagged population density, $\ln N_{t-d}$, the phase portrait narrowed at first but then widened again as the delay, d, was increased (Figure 9.6). One effect of this narrowing of the phase portrait is to increase the correlation between R and $\ln N_{t-d}$. For example, the correlations, r_d, between R and $\ln N_{t-d}$, $d = 1, 2, 3, 4$ for the larch budmoth time series are $r_1 = -0.38$, $r_2 = -0.78$, $r_3 = -0.79$, $r_4 = -0.5$. From this we concluded that the dominant feedback mechanism has a delay of $d = 3$. However, it is statistically possible for some of the correlation at lag 3 to be carried over from lags 1 and/or 2. These confounding effects can be removed by calculating the *partial correlation* between R and $\ln N_{t-d}$ with the effects of lower lags removed. In this way a *partial rate correlation function*[32], or PRCF, can be constructed for the time series (Figure 9.7).

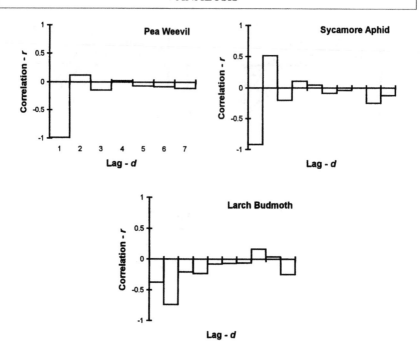

Figure 9.7 Partial rate correlation functions (PRCF) for stationary time series.

The PRCF is one of the most useful statistics for determining the order of the dynamics in stationary time series. For example, PRCFs created by first-order processes ($d = 1$) have large negative correlations at lag 1, like the pea weevil and sycamore aphid (Figure 9.7). Higher-order ̄feedback processes, on the other hand, have large negative correlation coefficients at longer lags (e.g. the larch budmoth PRCF in Figure 9.7). When large negative correlation coefficients are observed at more than one lag, it may indicate that the dynamics are affected by two or more feedback processes. In the case of the larch budmoth, for example, the PRCF suggests that the dynamics are governed by both first- and second-order feedbacks, but the second-order effect is the strongest (Figure 9.7). Notice that the third-order effect detected in the phase portrait (Figure 9.6) is not evident in the PRCF, indicating that this effect may be spurious.

9.3 NON-STATIONARY DATA

A major precondition for using most of the diagnostic tools presented above is that the time series is stationary. All is not lost, however, if the data are found to be non-stationary, for it may be possible to adjust them to fulfil the stationarity requirement.

The first thing to note is that time series data can only be made stationary if they exhibit some sign of equilibrium dynamics. In other words, there must be some reason to suspect that the population is fluctuating around equilibrium, or carrying capacity, even if that equilibrium changes with time (e.g. there is a trend or shift in the carrying capacity). Time series like those from the human and blue whale populations are obviously in a state of transition and do not exhibit any signs of equilibrium dynamics. Searching for stabilizing ⁻feedback in such data is meaningless.

9.3.1 Sequencing

When time series exhibit sudden changes in the mean, or *discontinuities*, they can sometimes be divided into two or more separate series at the point of discontinuity, an operation called *sequencing*. For example, the dynamics of the gypsy moth population changed dramatically around 1922 (Figure 9.2). A logical step would be to divide the time series into two sequences, a high-density sequence from 1911 to 1921 and a low-density sequence from 1922 to 1934. These two sequences can then be analysed as separate time series. If we do this for the gypsy moth time series, we will find that the first sequence (1911–1921) is also non-stationary, seeming to have a downward trend (we will analyse this sequence in the section on detrending below). On the other hand, the second sequence appears to be stationary and to be dominated by second-order (lag 2) feedback (Figure 9.8; Table 9.2).

The lake whitefish time series also exhibits a sudden change from stable dynamics prior to 1960 to a continuous decline thereafter. Dividing this time series into two sequences around this date enables us to evaluate the earlier sequence and to conclude that it was dominated by first-order (lag 1) feedback (Table 9.2). The second sequence shows a linear decrease on the logarithmic plot and could be treated as a transient to extinction. Another possibility, however, is that the dynamics followed a trend caused by a gradually deteriorating environment. We will examine this possibility below.

9.3.2 Detrending

If a non-stationary time series exhibits a constant trend over the period of observation, then the trend can sometimes be removed by rotating the series around a regression line through the data, an operation called *detrending*. For example, the first sequence of the gypsy moth series shows evidence of a linear trend on the arithmetic scale (Figure 9.9).

A trend such as that shown by the gypsy moth sequence could mean that the equilibrium, or carrying capacity, was decreasing as a linear function of time. If this was true, the appropriate trend model is

$$K_t = a + b(t - t_0), \tag{9.5}$$

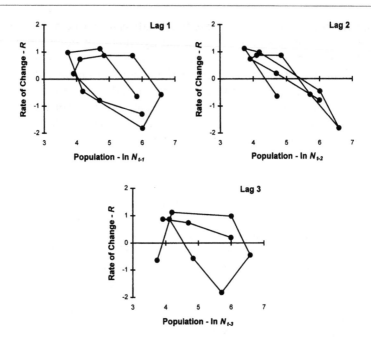

Figure 9.8 Phase portraits in time delay coordinates (R plotted on $\ln N_{t-d}$, $d = 1$, 2, 3) for the second gypsy moth sequence (1922–1934).

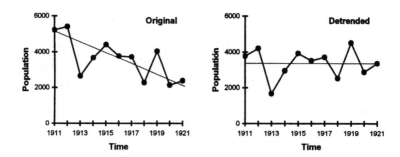

Figure 9.9 Detrending the first gypsy moth sequence; the original data are fitted to a linear regression line $K_t = 5045 - 241(t - 1911)$ with K_t the variable carrying capacity (line), and then the data are detrended to a common carrying capacity, K, by adding the difference between the original data and the regression line to the common $K = 3400$; i.e. $N_t' = N_t - K_t + 3400$, where N_t' is the new adjusted time series.

where a and b are regression constants[32], t is the current time, and t_0 is the time at the beginning of the trend (1911). After estimating the parameters of the model by regression, the series is detrended by adding the difference between the regression line and the data to a horizontal line, usually the mean of the series, using the equation

$$N'_t = N_t - [a + b(t - t_0)] + \overline{N}, \qquad (9.6)$$

where N'_t is the new (detrended) observation, N_t is the original observation, and $\overline{N}$ is the mean of the series (Figure 9.9). An analysis of the detrended first sequence leads to the conclusion that the dynamics of the dense gypsy moth population were dominated by first-order (lag 1) feedback. Remember that second-order (lag 2) dynamics were detected after the transition to sparse densities (Figure 9.8 and Table 9.2).

The second sequence of the lake whitefish data could also be caused by a trend in the lake environment rather than a transient towards extinction. This idea was tested by detrending the data and then examining the lag structure of the stationary sequence, and led to the conclusion that the order of the dominant feedback remained unchanged during the decline (Table 9.2). This conclusion suggests that the whitefish population was declining because of a deteriorating environment.

9.4 INTERPRETATION

Once the nature of the exogenous and endogenous factors governing the dynamics of a particular population have been diagnosed, it may be possible to erect hypotheses for the biological mechanisms responsible for generating the observed fluctuations. Here is where the principle of limiting factors can be useful, for it implies that one ‾feedback mechanism will probably dominate the dynamics, at least close to equilibrium. With this in mind, let us attempt to interpret the time series we have analysed in this chapter:

1. The *pea weevil* time series fluctuated around an apparent constant equilibrium density, or carrying capacity. Diagnosis indicated that the series was stationary and was dominated by first-order ‾feedback (lag 1). These observations lead to the hypothesis that *intra*specific competition for a non-reactive limiting factor (the third principle) was the probable mechanism regulating the dynamics of the pea weevil population. The most likely limiting factor is food because food was supplied at a fixed rate (non-reactive) and predators were not present in the environment. In addition, the R-function is concave, suggesting incidental *intra*specific competition. This conclusion is also reasonable because weevils are not expected to exhibit adapted (e.g. territorial) competition.

2. The non-stationary *human* population was observed to grow at a faster rate than predicted by the first principle over the period 1800 to the present. This implies that the second principle, intraspecific co-operation, may have dominated the dynamics during this period. As this time was marked by unprecedented advances in agriculture, medicine and industry, the logical conclusion is that these advances progressed as a direct function of human population density and resulted in increases in the per capita rate of change of the human population. The unstable dynamics of the human population may leave some with an uneasy feeling when they contemplate the apparent absence of the third and fourth principles and the possibility that they will be evoked in the future (see Chapter 16 for more on this issue).

3. The *blue whale* harvest data were also non-stationary and it seems likely that the geometric decline in catches was due to overexploitation causing the death rate (D) to exceed the birth rate (B). However, the rapid decline in harvests over the last years probably reflects the moritorium on blue whale hunting rather than a population on its way to extinction. In any respect, the time series does not seem to exhibit the hyper-exponential decline of a population below its extinction threshold, so the whale population should recover if left alone.

4. The *sycamore aphid* exhibited saw-tooth fluctuations around a more or less steady mean density, and the analysis strongly indicates that the dynamics are regulated by first-order (lag 1) ⁻feedback. However, it is not clear what the limiting factor(s) could be in this system. Many aphids are attacked by a suite of efficient generalist predators (e.g. hover fly larvae and ladybird beetles)[26], and the switching and aggregation of these predators in response to aphid density changes could give rise to the observed first-order dynamics. However, empirical evidence suggests that predators do not exert a strong influence on sycamore aphid populations. Aphids are also known to disperse, and thereby suffer high mortality, when their food plants become too crowded and this could also induce first-order dynamics. Intensive study of sycamore aphid dynamics[3] suggests that *intra*specific competition for food, and the resulting dispersal from overcrowded trees, creates the first-order feedback observed in the data.

5. The *larch budmoth* population went through three very regular cycles (9-year period) of enormous amplitude. Old records show that this cycle has repeated itself for at least 100 years. Diagnostics indicate that the dominant ⁻feedback operates with a lag of 2–3 years, or is a mixture of lag 1 and 2 feedback (this latter conclusion may be more reasonable because the budmoth cycle carries the population far from equilibrium and, thereby, violates the assumption under which the fifth principle holds). The current thinking is that larch budmoth dynamics

are generated by qualitative changes in the foliage of larch trees induced by heavy feeding in the previous 2 years[27].

6. For the first 11 years, the *gypsy moth* population exhibited predominantly first-order dynamics around a very high but declining mean density. The most plausible hypothesis is that its food supply, the leaves of its deciduous hosts, acted as a passive limiting factor (note that the leaves of deciduous trees are replaced every year and, therefore, can be considered as a non-reactive factor provided that the feeding of one gypsy moth generation does not influence the quantity and/or quality of next year's leaves). The gradual decline in the carrying capacity could be due to the suppression or death of preferred host trees (mainly oaks). The second, stationary, sequence of the gypsy series was dominated by second-order feedback, possibly due to the interaction with reactive specialist parasitoids that were introduced from Europe in an attempt to control the moth population[33] (the regulation of pest populations at innocuous levels by biological agents is known as *biological control*).

7. The *whitefish* population in Lake Ontario fluctuated with predominantly first-order dynamics around a more or less stationary mean for the first 23 years but then entered a period of continuous decline. Because the decline was also characterized by first-order dynamics, it seems logical to propose that the collapse was due to a deteriorating environment, perhaps overfishing and/or the incursion of the parasitic sea lamprey.

It is important to remember that diagnosis is an art, and that interpretations based on diagnostic analysis represent the opinion of the diagnostician. Others may disagree. In the case of the gypsy moth, for example, some scientists believe that small mammals rather than parasitoids limit gypsy moth populations at low densities[33]. Hence, diagnostic interpretations should be treated as hypotheses about the causes of population fluctuations and these hypotheses should be subjected to experimental tests before they are accepted as fact (see below). Caution is also required in interpreting the results obtained from the analysis of endogenous dynamics. In the case of the larch budmoth, for example, it is at least possible that the regular cycles of abundance were caused by a cyclical exogenous factor, such as, for instance, climatic cycles. At this time, however, no exogenous factor has been found that shows the rigid 9-year cycles observed in the budmoth population. For this reason, an endogenous explanation seems to be the most plausible. As a general rule, it is inadvisable to search for endogenous causation when there are a priori reasons to suspect that the dynamics are the result of exogenous factors. For example, 10-year cycles of abundance of the fir engraver beetle, an insect that kills fir trees in western North America, were found to be driven by episodes of drought and defoliation by another insect[34]. In addition, prior analysis had shown that fir engraver populations were limited by competition for food, a first-order process. Searching for endogenous causes of population cycles in this insect is, therefore, inappropriate and misleading.

9.4.1 Testing hypotheses

Whenever possible, diagnostic interpretations should be subjected to experimental testing. The most convincing test of an ecological hypothesis is to predict the effect of an experimental manipulation a priori (before the fact). For example, the pea weevil analysis led us to propose that *intra*specific competition for food (third principle) was the major determinant of the observed dynamics. This hypothesis is easily tested by repeating the experiment with a different food supply, say double the amount in the first experiment, because this should result in a proportional change in the equilibrium density if food is the limiting factor.

Hypotheses can also be tested by modelling the interaction between the population and the suspected limiting factor(s)[35]. For example, during the study of larch budmoth population dynamics (Figure 9.2), insect parasitoids were also censused. Hence, it was possible to build a model of the interaction between the budmoth and its larval parasitoids by fitting two-species R-functions to the data (see Chapter 10 for details). The low coefficients of determination ($r^2 = 0.568$ for the budmoth R-function and $r^2 = 0.676$ for the parasitoid R-function) suggest that the interaction with parasitoids alone is insufficient to explain the observed dynamics.

One of the best ways to test hypotheses for the mechanisms governing the dynamics of field populations is to perform *perturbation* experiments. Here the population is disturbed from equilibrium, by adding or removing organisms, and subsequent observations are made on changes in birth and death rates as well as parasites, predators and food supplies. A good example is an experiment in which large numbers of gypsy moth eggs were collected from outbreak regions and transplanted to areas with very sparse populations[14]. Within a season, all these local gypsy moth infestations had been reduced to almost zero by insect parasitoids, thereby supporting the parasitoid limitation hypothesis[33].

9.5 SUMMARY

In this chapter we

1. Defined the idea of ecological diagnosis.
2. Defined the concept of stationarity.
3. Defined the time series and the clues it holds to stationarity and order.

4. Defined the return time and its use for diagnosing stationarity and order.
5. Defined the autocorrelation function (ACF) and its use for detecting stationarity and period.
6. Defined the phase portrait and the clues it holds to order and stationarity.
7. Defined the partial rate correlation function (PRCF) and its use for detecting the order of population dynamics.
8. Described how to sequence and analyse time series that exhibit sudden discontinuities in their mean densities.
9. Described how to detrend and analyse time series that exhibit gradual changes in their mean densities.
10. Showed how to use the results of diagnosis to interpret the causes of population fluctuations.
11. Suggested how hypotheses concerning the causes of observed population dynamics can be tested by prediction, modelling and perturbation experiments.

9.6 EXERCISES

The data presented below show the estimated number of Soay sheep on the Island of Hirta[36] over the period 1961–1967: 910, 1056, 1590, 1006, 1469, 1598, 876.

1. Plot the time series*.
2. Calculate the mean and variance of the series.
3. Calculate the mean and variance of the return time.
4. Calculate the autocorrelation at lags 1, 2, and 3.
5. Plot the phase portrait.
6. Calculate the rate correlation at lags 1, 2, and 3.

*These exercises should be done by hand or on a computer spreadsheet.

<table>
<tr><td>10</td><td># Modelling:
predicting nature's rhythms</td></tr>
</table>

Here we arrive at a very interesting conclusion, namely, that the development of a definite biological system is conditioned not only by its state at a given moment, but that the past history of the system exerts a powerful influence together with its present state. (G. F. Gause, 1934)

The objectives of most ecological analyses are two-fold: first to obtain an understanding of the underlying cause or causes of observed phenomena (what has been called diagnosis), and second to forecast future changes in the system in the presence and absence of human intervention. To achieve the first requires diagnostic tools, the subject of the last chapter. To achieve the second requires *models*.

Models are conceptual representations of real phenomena which are usually, but not always, framed as mathematical equations (e.g. verbal or physical models). Mathematical models are preferable, however, because they present a precise and unambiguous statement of the concept (although not necessarily of the reality), so that the results obtained by one person are exactly the same as those obtained by another. In addition, mathematical models enable one to make *quantitative forecasts*.

10.1 THE BASIC FORECASTING EQUATION

All populations of living organisms obey, without exception, the first principle of population dynamics. In other words, they all have the potential to grow or decline geometrically. Hence, the basic element of any quantitative forecast of population change is the principle of exponential growth. Given estimates of the present state of a population, N_{t-1}, and of its per capita rate of change over a given interval of time, R, then its expected state at the end of the time interval, $E[N_t]$, is given by the familiar step-ahead forecasting equation

$$E[N_t] = N_{t-1}e^R. \tag{10.1}$$

The present state, or current density, of the population, N_{t-1}, can usually be estimated by some kind of sampling or census procedure, but an estimate of the realized per capita rate of change, R, may be more difficult to obtain. Thus, the most difficult problem in population modelling is to obtain a mathematical description of the R-function. It is important to realize that, unlike the first principle, which is identical for all populations, the R-function is unique to the species and may even be different for the same species living in different places or at different times.

10.2 QUANTITATIVE *R*-FUNCTIONS

Perhaps the best way to build quantitative R-functions is from data relating the per capita birth and death rates to current and/or past population densities. Unfortunately such data are difficult and expensive to obtain. Alternatively, R can be calculated from a time series through the relationship $R = \ln N_t - \ln N_{t-1}$, and the R-function estimated by regression of R on N_{t-d} (e.g. Figure 9.1)[37]. As time series data are often available to resource and pest management agencies, this is probably the most practical approach.

The first step in time series analysis and modelling is to determine if the series is *stationary* using clues in the return time and ACF (Chapter 9). If the series is stationary, then the endogenous R-function can be modelled by regression; if not, then the exogenous processes need to be modelled first.

10.2.1 Stationary time series

The first step in modelling a stationary time series is to determine the order of the endogenous dynamics using clues in the return time, phase portrait and PRCF, or any other available information (Chapter 9). Once this has been done, it is usually possible to fit the data to a particular R-function using standard *regression analysis*[32]. For example, the cowpea weevil data were fitted to the non-linear logistic model [equation (7.5)]

$$R = A - CN_{t-d}^Q, \tag{10.2}$$

with parameter values $A = 5.021$, $C = 0.8018$, $Q = 0.3536$ (see Figure 9.1). Remember that the order of the model, $d = 1$, was determined prior to model fitting. Also remember that the equilibrium density or carrying capacity, K, can be calculated from the relationship

$$K^Q = A/C$$

$$K = \exp\left(\frac{\ln A - \ln C}{Q}\right) = 179.13 \tag{10.3}$$

The *coefficient of determination*[32] is often used as a measure of the "good-ness of fit" of model to data. In the case of the cowpea weevil regression, the coefficient of determination was $r^2 = 0.995$ which implies that the modelled R-function explains 99.5% of the variation in pea weevil population change. It also implies that 99.5% of the observed fluctuations were caused by endogenous feedback and only 0.5% by exogenous random effects. Because of the good fit, the model would be expected to provide accurate quantita-tive forecasts of cowpea weevil density changes (but see below).

On occasion, time series data may also be available for another species or group of species inhabiting the same environment as the one we want to forecast. For example, during the sampling of larch budmoth popula-tions (Figure 9.2), the number of insect parasitoids was also counted. Hence, it was possible to estimate the parameters of a two-species R-func-tion using equation (6.5) (note that we would have used equation (6.4) if the other species was a resource rather than an enemy). Fitting the two-species logistic model

$$R_N = A_N - C_N N_{t-1} - C_{NP}\frac{P_{t-1}}{N_{t-1} + F}, \tag{10.4}$$

to the budmoth–parasitoid time series by multiple regression analysis provided the following estimates of the budmoth parameters: $A_N = 2.67$, $C_N = 0.0001$, $C_{NP} = 4.812$, $F = 0$, $r^2 = 0.568$ (the alternative prey para-meter F was assumed to be zero because no information was available on the availability of alternative prey). Using this model, the rate of change of the budmoth can be calculated from current estimates of budmoth and parasitoid densities. However, the low coefficient of determination ($r^2 = 0.568$) suggests that parasitoids were not playing a dominant role in the dynamics of larch budmoth populations.

The mathematical models developed in this book and discussed above are mainly variants or extensions of the *logistic* equation, an equation that is familiar to most ecologists and which also has a logical theoretical foundation[16, 17]. However, realizing that models are only abstractions of reality, it is entirely possible that other models may, on occasion, provide better fits and more reliable forecasts. Three alternative models are described below:

1. *The Gompertz equation.* One alternative is to consider R to be a func-tion of the natural logarithm of population density

$$R = A - C\ln N_{t-d}^Q, \tag{10.5}$$

which is known as the Gompertz equation.

2. *Multiple time delays.* Another alternative is to use models with multiple time lags. Remember that the concept of limiting factors (the fifth principle) was employed to simplify the structure of the logistic model (Chapter 7), but that this principle only applies strictly in the vicinity of a stabilizing equilibrium. When observed population densities fluctuate far from equilibrium, which is particularly evident in the presence of delayed feedback (fourth principle), it may be advisable to use multiple feedback models such as equation (7.4); i.e.

$$R = A - C_1 N_{t-1}^{Q_1} - C_2 N_{t-2}^{Q_2} - \cdots C_d N_{t-d}^{Q_d}, \tag{10.6}$$

where d is now the maximum time delay or dimension of the system (Chapter 7). Obviously, we can also have multiple delay Gompertz R-functions when the independent variables are transformed to logarithms[38]. Multiple delay models can be fitted to time series data with multiple regression procedures.

3. *Lotka–Volterra models.* There are also alternative two-species models, the best known being the "Lotka–Volterra"[23]. Here, the effect of the other species is proportional to its density rather than the consumer/resource ratio in logistic models. For example, a discrete-time version of the Lotka–Volterra predator–prey model with fixed and varying resources is

$$\begin{aligned} R_N &= A_N - C_N N_{t-1} - C_{NP} P_{t-1} \\ R_P &= C_{PN} P_{t-1} - D_P - C_P P_{t-1}, \end{aligned} \tag{10.7}$$

where N and P are the densities of the prey and predator species, respectively, and D_P is the death rate of starved predators. The Lotka–Volterra model can also be fitted to time series data by multiple regression and may be more applicable in situations where consumers do not compete for reactive resources or where their enemies are insatiable. In the case of the larch budmoth and parasitoid time series, the two-species Lotka–Volterra model only explained 28% of the variation in the budmoth per capita rate of change. However, there are situations (e.g. some harvesting methods that do not saturate) where the Lotka–Volterra model produces better fits than the logistic, and so the equation is a useful addition to the modeller's toolbox.

Finally, we should be aware that the approach taken in this book neglects a whole family of models that incorporate age structure into their framework. This is because population census data are usually made up of a single age class, or an unspecified mixture of age classes, so that it is impossible to reconstruct the age structure of the population. When it is necessary to consider the age and/or size distributions, other census methods, analytical techniques, and modelling approaches must be employed.

10.2.2 Non-stationary time series

When time series are found to be non-stationary, additional models may be required to account for the effects of exogenous factors. There are four major kinds of non-stationary time series:

1. *Transients.* Time series exhibiting transient dynamics are usually under the influence of the first principle alone, or sometimes by the combined action of the first and second principles. For example, the human population of the world seems, at first glance, to be growing exponentially as expected under the first principle. If this is true, then the R-function should have a slope of zero (no feedback), R will be constant, and so the R-function does not need to be modelled. On the other hand, if the second principle is acting on the human population, then it will grow faster than exponentially (*hyper-exponential* growth) and R will have a positive relationship to population density. This is apparently the case with the human population (Figure 10.1), and so a model of the human R-function is required for forecasting future population numbers.

2. *Trends.* If the time series exhibits a trend in the mean, then the trend must be modelled before reliable forecasts can be made. For example, the harvest of whitefish from Lake Ontario declined continuously over the last 10 years of the time series (Figure 9.2). Obviously, if forecasts are to be made of whitefish catches after this time, the trend must be

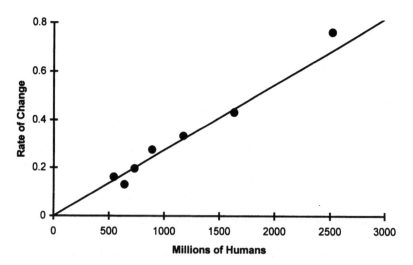

Figure 10.1 Per capita rate of change of the world human population estimated every 50 years[39] from 1650 to 1950 and plotted against the initial size of the population. The solid straight line is a simple linear regression through the origin $R = C N_{t-1}$ with $C = 0.000273$, $r^2 = 0.95$.

assumed to continue. Because the trend seems to have been due to exogenous forcing (see Chapter 9), and as the decline was roughly linear on the logarithmic scale, it can be modelled by a linear regression equation [equation (9.5)]

$$\ln K = 4.907 - 0.266(t - 1962), \tag{10.8}$$

where K is the equilibrium density (carrying capacity), and $t - 1962$ is the time (in years) elapsed since the trend started (1962). It is now possible to detrend the data [equation (9.6)] and then to analyse its endogenous dynamics. This analysis indicates that the order of whitefish population fluctuations was similar before and after the trend started, suggesting that the same endogenous mechanism was operating over the entire time series (Chapter 9). Because of this, it makes sense to join the first stationary sequence to the second detrended sequence and so create a longer time series for modelling the *R*-function. This is done by adjusting the mean of the detrended sequence to that of the stationary sequence by subtracting the difference between the two means from the detrended data (Figure 10.2). The whole time series can now be fitted to the data to estimate the *R*-function. Notice that this model, a first-order logistic equation, only explains 47% of the variation in the data, which means that more than 50% of the variation is due to exogenous random environmental variability (Figure 10.2). The expected whitefish harvest in 1973 can now be forecast by calculating the expected carrying capacity in 1972 from the trend model [equation (10.9)]

$$E[K] = e^{4.907 - 0.266(1972 - 1962)} = 9.46;$$

using this value to calculate the coefficient of competition, $C = A/K = 0.889/9.46 = 0.09397$; calculating the expected per capita rate of change in whitefish harvest with equation (10.2) (see Figure 10.2 for parameter estimates) given a 1972 harvest of 10 metric tonnes

$$E[R] = 0.889 - 0.09397 \times 10 = -0.0507;$$

and then computing the harvest in 1973 with the forecasting equation (3.12)

$$E[N_{1973}] = 10 \times e^{-0.0507} = 9.5 \text{ metric tonnes.}$$

3. *Discontinuities.* Non-stationary time series can also arise if a sudden change in the environment causes a shift in the mean of the series. For example, the gypsy moth population suddenly declined in the early 1920s from a high density of around 5000 egg masses per acre to a low density of around 200 egg masses per acre (Figure 9.2). Analysis of the two sequences showed that the population dynamics changed from predominantly first order to predominantly second order at about

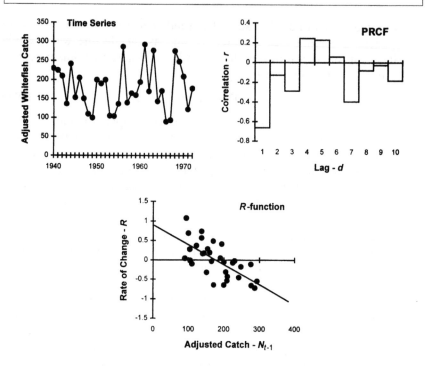

Figure 10.2 Lake whitefish time series with the last 10 years detrended and added to the end of the previous 23 years using the transformation $\ln N_t' = \ln N_t - [4.907 - 0.266(t - 1962)] + 5.124$, where N_t' is the adjusted time series from 1962 to 1972 and 5.124 is the logarithm of carrying capacity K of the series from 1940 to 1961; the partial rate correlation function (PRCF) for the adjusted series; and the fit of the single-lag linear logistic R-function $R = A - CN_{t-1}Q$ to the data with parameter estimates $A = 0.889$, $C = 0.0051$, $K = 175$, $Q = 1$, $s = 0.25$, and $r^2 = 0.473$.

the same time (Figures 9.5 and 9.8). This observation precludes the use of the first sequence for forecasting future gypsy moth population changes (but see *Multiple equilibria* below). Thus, to forecast gypsy moth densities after 1934 the last 13-year sequence is fitted to a non-linear, second-order R-function [equation (10.2)]

$$E[R] = 2.25 - 0.3034 N_{t-2}^{0.387}$$

with coefficient of determination $r^2 = 0.81$ and standard deviation $s = 0.417$.

4. *Multiple equilibria.* An alternative hypothesis for the discontinuity in the gypsy moth time series is that the second principle (escape from natural enemies) dominates over intermediate population densities[33]. According to this hypothesis, gypsy moth populations can escape from regulation at sparse densities and rise to very high (outbreak) levels.

Modelling this system would require two R-functions (Figure 7.4) because the population can now switch from low- to high-density dynamics and vice versa following random environmental disturbances. An estimate of the unstable escape threshold is also required. However, because natural populations are rarely observed near their unstable equilibria, data are usually unavailable and so indirect methods, as outlined below, must be used to estimate this threshold.

If an unstable threshold exists, there will be a line dividing N-R phase space into two basins of attraction, one to the low-density equilibrium (J) and another to the high-density equilibrium (K). This line is called a *separatrix* and is defined by

$$R_S = \ln(E/N_{t-1}) = \ln E - \ln N_{t-1}, \tag{10.9}$$

where R_S is the value of R on the separatrix and E is the constant parameter, the escape threshold, that we want to estimate. Because equation (10.9) defines a straight line with a slope of unity, its approximate location can be found from the logarithmic phase portrait; i.e. the plot of R on $\ln N_{t-1}$. The separatrix will be a straight line with a slope of unity that divides the data into two groups, a high-density sequence and a low-density sequence (Figure 10.3). By adjusting this line until it cleanly segregates the data into two groups, a rough estimate of the escape threshold is obtained by noting the point where the separatrix crosses the $R = 0$ line. Notice that all the vectors to the right of the gypsy moth separatrix converge towards the high-density equilibrium (first-order dynamics), while those to the left cycle around the low-density equilibrium (second-order dynamics). The vector that crosses the separatrix is assumed to be the result of random exogenous factors that displace the trajectory across the separatrix and initiate the switch from high- to low-density dynamics.

Using this approach, the gypsy moth escape threshold was estimated to be $E = 1212$ (Figure 10.3), and the two-equilibrium R-function is

$$E[R] = \begin{cases} 1.46 - 0.0003056N_{t-1}, \ N_{t-1} > 1212 \ldots \text{high-density} \\ 2.25 - 0.3033N_{t-2}^{0.387} \ N_{t-1} < 1212 \ldots \text{low-density} \end{cases} \tag{10.10}$$

10.3 QUALITATIVE R-FUNCTIONS

The construction of quantitative R-functions requires time series data collected over a relatively long period of time. However, even when such data are not available, it may still be possible to estimate the R-function from knowledge about the biology and ecology of the species and/or independent estimates of its parameters. For example, the maximum per capita

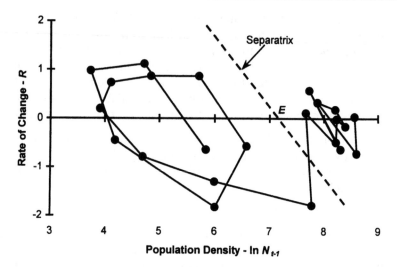

Figure 10.3 Logarithmic phase portrait of the gypsy moth time series with an approximation of the separatrix that divides the data into high-density and low-density basins of attraction; E is the estimated escape threshold.

rate of change (A) could be estimated from field data collected during periods when population density is very sparse and the third principle is not operating; i.e. $A = \max R = \ln(1 + B_{max} - D_{min})^7$. Likewise, historical information can provide information on the average density of the population, and this information can be used to estimate the equilibrium densities, J and/or K. Because models constructed in this way usually require some degree of qualitative assessment, or even guesswork, they are referred to as *qualitative models*.

It is sometimes possible to build useful qualitative models without any hard data at all. For instance, the general shape of the R-function for a particular population can sometimes be deduced from an understanding of the principles of population dynamics and knowledge of the biology and ecology of the organism. Knowing that the dominance of the third principle (*intra*specific competition) causes the R-function to decline (negative slope) with population density, while the dominance of the second (*intra*specific co-operation) causes the slope to rise with population density (positive slope), the basic form of the R-function could possibly be deduced from knowledge about how the relative dominance of these two interactions changes with population density. For example, the second principle is expected to dominate at very sparse densities because competition for resources should be weak while the effects of co-operation for mates should be strong, at least in sexually reproducing species. Similarly, the third principle is expected to dominate at very high densities because *intra*specific competition for limiting resources should be intense. Therefore, the *R*-functions for many populations will have a positive slope at very low densities and a negative slope at very high densities (see, e.g. Figure 7.2). If the main interest is in forecasting changes in relatively large populations, then the underpopulation effect can be ignored, and a single *R*-function with a negative slope will often suffice. This *R*-function will have a single equilibrium (K) that can sometimes be estimated from data on the abundance of a limiting resource, such as food and/or space, or from observation of the average population density.

The next thing to consider is the possibility of non-linearity in the *R*-function. For instance, if it is known that competition for limiting resources involves territorial or other aggressive social behaviour (adapted competition), then it is likely that the slope of the *R*-function will become steeper with population density (down-curved or convex; $Q > 1$). On the other hand, if competition is non-adapted, the slope will probably decrease with density (up-curved or concave; $Q < 1$).

The possibility of complex *R*-functions must also be considered. For example, the population may be limited to densities well below its potential carrying capacity by enemies or an inability to utilize potential food supplies that are resistant to attack. Under these conditions, *intra*specific competition for enemy-free space or susceptible food can create a second low-density *R*-function. This kind of *R*-function requires that the second principle dominates at intermediate densities, and so is most likely to be found in species that co-operate during defence or attack (adapted group hunting or defence behaviours) or in those limited by non-reactive generalist predators that switch and/or aggregate in response to prey density (incidental co-operation). Complex *R*-functions should be suspected in populations with these characteristics.

Once the basic form of the *R*-function for a particular population has been deduced, the next thing to consider is the possibility that the fourth principle (circular causality) may be involved in the dynamics around equilibrium. This requires knowledge about the reaction of limiting factors to the density of the population. For instance, if the population is limited by enemies, then the fourth principle may be evoked if enemies reproduce more offspring when their food supplies (the prey population) increase; i.e. if the enemies have a strong numerical response to prey density. This is more likely to occur if enemies are *specialists*, feeding exclusively on the species of interest. On the other hand, when the population is limited by food or other resources, the fourth principle is likely to be evoked if high population densities dramatically reduced the abundance or quality of the resource.

The above considerations lead to a general strategy for deducing the form of the qualitative *R*-function for a population:

1. Determine the limiting factor or factors, remembering that a limiting factor may consist of a group of co-acting environmental components, like a guild of predators.
2. Determine if there is the potential to escape from the constraints imposed by one or more of the limiting factors and, therefore, if more than one R-function needs to be considered.
3. Determine for each R-function if the limiting factor is reactive or non-reactive and, from this, if time delays are likely to be encountered in the feedback process.
4. Determine if non-linearities in the R-function should be considered.
5. Attempt to obtain estimates of the parameters of the R-function.

Examples of qualitative modelling will be encountered in Chapters 14 and 15.

10.4 FORECASTING

There are two basic approaches to forecasting population changes: the first attempts to estimate the actual density of organisms expected in the future, or the range of densities expected with a certain degree of confidence. This is called a *point forecast*, and the probability that the forecast will fall within a given range of densities is called the *confidence interval* of the point forecast. The second approach attempts to forecast the *probability* of the population changing in a particular direction (e.g. increasing or decreasing) or of it exceeding a particular threshold level (e.g. the "outbreak threshold" or "economic damage level" in the case of a pest species, and the "danger of extinction" in the case of an endangered species). We call this a *probability forecast*.

10.4.1 Point forecasts

Point forecasts are made by calculating the expected per capita rate of change from an R-function and then inserting this value into the step-ahead forecasting equation (10.1) to calculate the expected density of the population at the end of a period of time, usually a year. For example, the density of cowpea weevils in generation 11, N_{11}, can be forecast from its known density in generation 10 ($N_{10} = 180$) by first calculating the expected R-value, $E[R]$, from equation (10.2) using the parameter values estimated from the time series by linear regression (Figure 9.1)

$$E[R] = A - CN_{10}^Q = 5.021 - 0.8018 \times 180^{0.3535} = -0.0086.$$

This value is then inserted into the step-ahead forecasting equation (10.1) to calculate the expected density in generation 11

$$E[N_{11}] = N_{10}e^R = 180e^{-0.0086} = 178.46.$$

Confidence intervals can be placed around the expected value as follows: knowing that 68% of the forecasts will fall within one standard deviation of the mean, the 68% confidence interval is given by

$$E[N_t] = N_{t-1}e^{R \pm s}, \qquad (10.11)$$

where s is the calculated standard deviation of the data around the fitted R-function[32]. For the cowpea weevil data, $s = 0.0774$ and the 68% confidence interval is

$$\text{Upper limit} = 180e^{-0.0083 + 0.0774} = 192.88,$$

$$\text{Lower limit} = 180e^{-0.0083 - 0.0774} = 165.22.$$

What this confidence interval means is that there is a 68% chance of being correct when we forecast the cowpea weevil density in generation 11 as being between 165.22 and 192.88. Increasing the interval provides forecasts with a higher degree of confidence. For example, the 95% confidence interval is 1.96 times as large so that

$$E[N_t] = N_{t-1}e^{R \pm 1.96s}, \qquad (10.12)$$

which turns out to be 153.38–207.76 for the cowpea weevil.

There is a slight problem with this method of calculating confidence intervals because the non-linear models being used do not conform strictly to statistical theory. An alternative method is to make a large number of forecasts, say 1000, with the stochastic version of the step-ahead forecasting equation

$$E[N_t] = N_{t-1}e^{R \pm sZ}, \qquad (10.13)$$

where Z is a variate chosen at random from a standard normal distribution (Appendix). The point forecast is then the mean of these 1000 stochastic forecasts, and the 95% confidence interval is obtained by deleting 2.5% of the largest and smallest values (i.e. deleting the 25 largest and the 25 smallest values from the 1000 forecasts). This method of calculating statistics from the model is known as *bootstrapping*. In the case of the cowpea weevil, a bootstrapped point forecast of $E[N_t] = 178.9$ was obtained with a 95% confidence interval of 152.4–205.4. In this particular example, the bootstrapped estimates are very similar to those obtained by the standard method.

10.4.2 Probability forecasts

Rather than trying to forecast the actual number of organisms in a population, probability forecasts estimate the probability of an event occurring. The event may be an increase or decrease in density, or the chance that the population will exceed a particular threshold density (e.g. an outbreak or extinction threshold). For example, the gypsy moth model [equation (10.10)] can be used to forecast the probability of an outbreak starting or ending. In this case the stochastic equation (10.13) is used to make 1000 or so predictions and the number of instances in which the predicted population density crossed the escape threshold is noted (remember that each prediction will be different because different random numbers are used for each stochastic forecast). The proportion of predictions that cross the escape threshold is then used as an estimate of the probability of outbreak or collapse. Using this approach, the probability of the gypsy moth population reaching outbreak levels in 1935 was calculated to be zero.

10.5 FORECASTING WITH CONFIDENCE

Confidence in forecasting rests, first and foremost, on a personal conviction that the theory of population dynamics is a logical and reasonable representation of reality, and that the model being used is a reasonable representation of that theory. Of course, confidence also depends on the data being used to estimate model parameters for, if one believes that the data truely reflect the real dynamics of the population, then confidence in forecasts based on these data will be higher. Apart from this, confidence is also increased if the model behaves in a similar way to the true population and, in particular, if its forecasts are proved by time to be correct. These latter factors are amenable to objective testing.

10.5.1 Testing model behaviour

The dynamic behaviour of a model is usually studied by *simulation,* or the calculation of model trajectories through time. Simulations can be run in the absence of external variability, in which case they are called *deterministic* or *steady-state* simulations, or in the presence of random environmental disturbances, in which case they are called *stochastic* simulations. Deterministic simulations allow one to see how the population would behave in an invariant environment; i.e. whether the equilibrium points are stable and, if not, whether the dynamics are periodic or chaotic (see Chapter 5). For example, deterministic simulations with the cowpea weevil model show it to be damped-stable, while the larch budmoth model produces a stable cycle (Figure 10.4).

Because real data are normally affected by exogenous disturbances, it is usually better to compare them to stochastic simulations. Notice that stochastic simulations with the pea weevil and larch budmoth models behave in a manner consistent with the original time series (e.g. they have the correct periodicity). However, because a different sequence of random

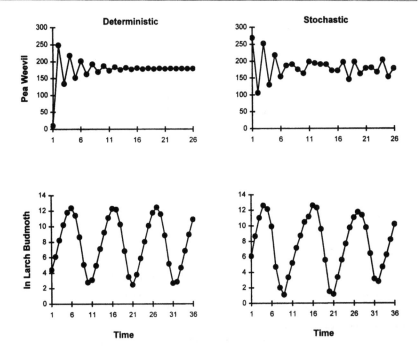

Figure 10.4 Deterministic (left) and stochastic (right) simulations with a non-linear first-order logistic model of the cowpea weevil data [equation (10.2); see Figure 9.1 for parameters] and a non-linear third-order Gompertz model of the larch budmoth data [equation (10.5) with parameters $A = 2.086$, $C = 0.00168$, $Q = 3.227$, $d = 3$, $r^2 = 0.71$].

disturbances is used in each simulation, the simulated data cannot be compared directly to the original time series. They can, however, be compared in phase space where the time dimension is invisible (Figure 10.5). Notice, that the stochastic simulations superimpose quite well on the original data, indicating that the behaviour of the models is similar to the real systems.

The procedure for comparing simulated and real data in phase space may involve adjusting the standard deviation to obtain coincidence between data and simulated trajectory. This is called *tuning* the model. Notice that the model should not be tuned by adjusting its parameters as this would introduce bias into the analysis; for example, one can make a model do anything one wants to – from stable equilibrium to chaos – by tuning its parameters (see Chapter 5). Adjusting the standard deviation, on the other hand, only alters the degree of random disturbance, not the basic behaviour from the model. In most cases, the standard deviation will need to be reduced from the value estimated by fitting the model because some of this variability reflects sampling error and other non-environmental effects.

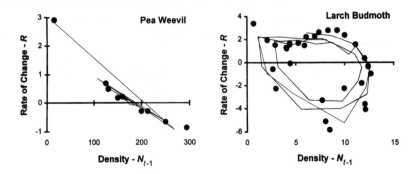

Figure 10.5 Phase portrait for stochastic simulations (thin line) with the cowpea weevil and larch budmoth models superimposed on the original data (dots).

10.5.2 Testing forecasts

Confidence in a model will also depend, for obvious reasons, on the accuracy of its forecasts. Confidence building of this kind, however, usually has to await the future. In the meantime, it may be possible to test forecasts against independent data from the same or similar populations. Even if such data are not available, the model can be tested against the original series by omitting data points.

Naturally, the best test of a forecast is against independent data. In the case of the cowpea weevil, for example, several different experiments were run under identical environmental conditions. Using the cowpea weevil model, with parameters estimated from the first time series, the expected densities of a second experiment could be predicted and compared to the actual data (Table 10.1). Notice that even with this very good model, the actual data points only fell within the 95% confidence intervals of the point forecasts in less than half the cases. On the other hand, probability forecasts were correct on all occasions.

Even if independent data are not available, it is still possible to test the model by omitting the last datum, or any other data point. The remaining data are then used to build the forecasting model, and the forecasts are tested against the omitted datum. When this is done with some of the time series discussed in this chapter, only the pea weevil model provides an accurate quantitative forecast of the final data point (predicted 95% confidence interval 179.6–228.9 versus the actual data 180). The general conclusion is that it may be unrealistic to expect precise point forecasts of field populations. On the other hand, accurate probability forecasts can generally be obtained with most of the models (Table 10.2). In fact, the only incorrect probability forecast was with the gypsy moth model, but even in this case the prediction that the population would not attain outbreak levels was correct.

Table 10.1 Comparison of cowpea weevil numbers forecast by model (10.2) (see Figure 9.1 for parameters estimated from Utida's Figure 1a) compared with data from Utida's Figure 1c: mean and 95% confidence intervals in each row were calculated from the observed density in the preceding row

Time	Observed	Predicted	95% interval	P(increase)
1	16	–	–	–
2	330	285.7	243.6–327.8	1.0
3	150	98.8	85.4–112.2	0.0
4	250	204.4	172.4–236.4	1.0
5	151	132.4	114.2–150.6	0.0
6	205	202.9	174.0–230.1	1.0
7	162	160.6	137.0–184.2	0.0
8	204	194.1	166.7–121.5	0.99
9	152	162.2	137.0–187.2	0.0
10	175	202.7	171.6–233.8	1.0

Table 10.2 Predicted probabilities of population increase from models fitted to data described in the text. The last datum of the time series was not used for parameter estimation but instead was used to verify forecasts

Species	Model	P(increase)	Observed
Pea weevil	Non-linear first-order logistic	1.0	Increase
Sycamore aphid	Non-linear first-order logistic	0.17	Decrease
Larch budmoth	Non-linear third-order Gompertz	0.95	Increase
Gypsy moth	Multiple equilibrium logistic	0.98 (0.0*)	Decrease
Lake whitefish	Non-linear first-order logistic	1.00	Increase

* Probability of exceeding the escape threshold $E = 1212$.

10.5.3 Testing prescriptions

Models are also useful for designing and testing management strategies and prescriptions. For example, managers of fisheries or other wildlife populations are often interested in maximizing the harvest of these natural resources, while pest managers are concerned with minimizing the probability of pest damage.

10.5.3.1 Harvesting resources

The problem of harvesting resources is often phrased as *maximizing the sustainable yield*. The sustainable yield from a population of organisms is the number (or biomass) that can be harvested in perpetuity from a population of given size. This is sometimes called the *surplus production* of the population, and is defined mathematically by $Y = N_t - N_h$, where N_t is the number of organisms present in the population when it was harvested to a density of N_h in the previous year. From this it follows that

$$Y = N_t - N_h = N_h e^R - N_h = N_h(e^R - 1). \tag{10.14}$$

Given a model for the R-function, then the sustainable yield curve can be calculated from equation (10.14) (Figure 10.6, left). Notice that the yield curve for the cowpea weevil reaches a maximum at a fairly low density of $N_{h,max} = 27$. This means that around 280 weevils can be harvested each generation if 27 are left to reproduce the next generation. The theoretical maximum sustainable yield of pea weevils is, therefore, $Y_{max} = 280$ per generation.

In real life, of course, the sustainable harvest is affected by random environmental disturbances. In Figure 10.6 (right) this effect is simulated by running the cowpea weevil model for 300 generations in a randomly varying environment for many different levels of harvest. For each experiment, a constant number of pea weevil adults is harvested at the end of each generation, before they can reproduce, but no more that 99% of the population is harvested and no harvest is allowed in any generation that the adult weevil population drops below 27. The results indicate that harvesting at the theoretical maximum sustainable yield (280 weevils per generation) is not the best strategy. Not only is the average yield lower in a variable environment, but about 10% of the time harvests have to be cancelled because the adult population is too low. The best strategy in a variable environment is to harvest around 200 weevils per generation or, if the objective is to minimize the number of "no harvests", then a better strategy may be to harvest around 150–160. The general lesson is that, in real life, the optimal harvesting strategy is usually below the theoretical maximum sustainable yield.

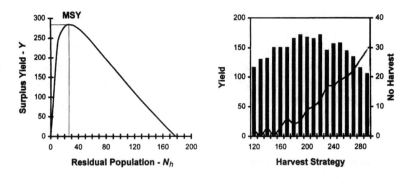

Figure 10.6 Sustained yield curve for the cowpea weevil model showing a maximum sustainable yield of $Y = 284$ at population density $N_h = 27$ (left), and the results of 300 years stochastic simulations with the same model for different maximum harvesting strategies; the average yields from each harvesting strategy are shown in the histogram and the number of years when no harvests were possible is shown as a line (right).

10.5.3.2 Managing pests

Pest management is similar to resource harvesting as the ultimate goal is maximization of the yield from a resource production system. However, the system now contains other organisms, the pests, which also harvest the resource. The problem is to maximize yields in the face of competition from pests.

The pest management problem is often defined in terms of an *economic injury level* or EIL, the pest density that causes sufficient damage to justify management action; i.e. the economic loss due to the pest is equal to the cost of controlling the pest. The decision to control is usually made at some pest density below EIL, called the *economic threshold* (ET) or *control action threshold* (CAT)[40]. The integration of pest management into the resource production system complicates the problem of maximizing yields because the interactions between three trophic levels – crop, pests and enemies – have to be considered. Some examples will be encountered later in this book.

10.6 SUMMARY

In this chapter we learned:

1. How to use the step-ahead forecasting equation.
2. How to fit models of the *R*-function to stationary time series data using regression analysis.
3. Some of the different models that can be employed to describe the *R*-function.
4. How to model transients, trends and discontinuities.
5. How to estimate the escape threshold in multiple equilibrium models.
6. How to use models to make point and probability forecasts.
7. How to estimate confidence intervals.
8. How to bootstrap point forecasts and confidence intervals.
9. How to construct qualitative models.
10. How to test the behaviour of population models.
11. How to test forecasts.
12. How to test management prescriptions.

10.7 EXERCISES

1. Build a model for the cowpea weevil time series* (Table 10.2) and test its predictions against the following independent time series: 16, 330, 150, 250, 151, 205, 162, 204, 152, 175.

2. Build a model for the gypsy moth population* described by the egg masses per acre time series (1911–1934): 5214, 5407, 2635, 3658, 4400, 3751, 3702, 2273, 4032, 2134, 2387, 402, 110, 50, 61, 127, 303, 722, 407, 66, 42, 112, 344, 181.

3. Use the following estimates of the human population numbers from 1650 to 1950 to calculate the 50-year per capita rate of change and to build a model† of the human R-function (numbers in millions): 545, 641, 731, 890, 1171, 1633, 2516. Use the model to predict the human population in the year 2000. How does this compare with current estimates of the human population? How many humans will there be in the year 2050 according to this model? How does this compare with other models[39]? What does this prediction mean to you?

*Models can be constructed with standard statistical software packages that can perform non-linear regression, or with the program *P1a – Single-species Time Series* in *PAS* mentioned in the Preface.
†The best way to build this model is with a spreadsheet.

PART THREE

Practise

11

Sandhill cranes: territorial behaviour?

Indians and other natives take cranes in subsistence hunting; Canadian, American and Mexican hunters take them for sport. Deciding who gets how many will be a challenging task, particularly as demands grow and if crane populations dwindle. (D. H. Johnson, 1979)

In this section of the book, the principles and tools learned in the previous sections are used to study the dynamics of real populations. In the first example, the methods are used to study the population dynamics of sandhill cranes (*Grus canadensis*) on the mid-continental flyway, a causal mechanism for the observed population fluctuations is proposed, and the resulting model is used to study the management of this species.

Sandhill cranes spend the winter months in the southern United States and northern Mexico and summers on their breeding grounds in the northern United States and Canada. Crane populations have been censused annually since 1959 along the Platte River in Nebraska, a traditional staging area for 79–99% of the cranes during the northward spring migration[41].

11.1 ANALYSIS OF SANDHILL CRANE DYNAMICS

The sandhill crane time series and *ACF* are shown in Figure 11.1. Visual inspection of the time series and *ACF*, and the calculation of return time statistics ($MRT = 3.122$; $VRT = 16.002$), indicate that the time series is non-stationary, probably because the average numbers of cranes increased by about 50,000 birds in the middle of the series. For this reason, the data were divided into two equal sequences (shown by the two lines in Figure 11.1).

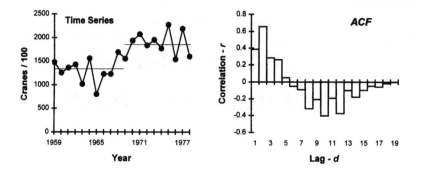

Figure 11.1 Time series data and autocorrelation function (ACF) for the sand-hill crane population censused each spring along the Central Flyway in Nebrasksa[41] (numbers of cranes divided by 100 for convenience). The two thin horizontal lines identify two 10-year sequences with different average densities.

11.1.1 Analysis of the first sequence (1959–1968)

Diagnostics from the first 10-year sequence are presented in Figure 11.2. The sharp saw-toothed oscillations of the time series, *ACF* period 2, narrow phase portrait, and dominant *PRCF* at lag 1 indicate first-order dynamics. Fitting a first-order linear logistic *R*-function to the data yields the statistics $A = 1.704$, $K = 1274.6$, $Q = 1$, $r^2 = 0.754$ (Figure 11.3). However, the value for the maximum per capita rate of change ($A = 1.704$) is biologically unreasonable because it implies a maximum reproductive rate of around $B = e^{1.704} - 1 = 4.5$ per bird per year, or nine offspring per pair [calculated from equation (2.10) with the death rate set to zero]. In an attempt to correct this, a non-linear logistic *R*-function was fitted to the data with the parameters $A = 0.466$, $K = 1343.74$, $Q = 5.638$. This model not only improved the fit ($r^2 = 0.889$) but also gave a more reasonable estimate of the maximum reproductive rate of 1.2 offspring per year per pair. The coefficient of determination was not improved by any of the alternative models.

Deterministic simulations indicated that the equilibrium point of the non-linear logistic is unstable, a conclusion that can be derived analytically from the stability criterion $AQd = 2.63$ (remember from Chapter 7 that the equilibrium point is unstable if $AQd > 2$). The dynamics are characterized by a stable oscillation that repeats itself every four time steps; i.e. complex dynamics with period 4 (Figure 11.3). Stochastic simulations mimicked the typical sharp saw-toothed oscillations seen in the data and corresponded well with the original data on the phase portrait, verifying that the model behaved in a manner consistent with the real system (Figure 11.3).

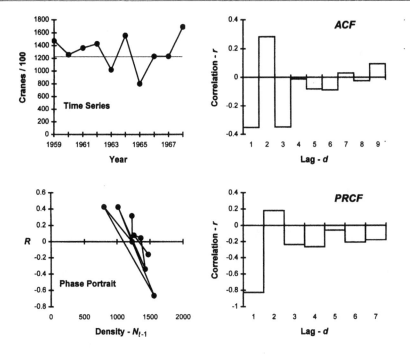

Figure 11.2   Diagnostics from the first sandhill crane sequence (1959–1968).

11.1.2 Analysis of the second sequence (1969–1978)

Diagnostics from the second 10-year sequence are presented in Figure 11.4. Again, the sharp saw-toothed oscillations, *ACF* period 2, narrow phase portrait, and dominant *PRCF* at lag 1 indicate first-order dynamics similar to those of the first sequence. Fitting a first-order logistic *R*-function to the data yielded similar parameters to the first sequence ($A = 1.802$, $K = 1906.46$, $Q = 1$, $r^2 = 0.912$ for the linear model and $A = 0.522$, $K = 1948.27$, $Q = 3.749$, $r^2 = 0.939$ for the non-linear logistic) (Figure 11.5). Notice, however, that the equilibrium density or carrying capacity was considerably higher in the second sequence, and that the model fitted the data even better (higher r^2 values).

Deterministic simulations showed that the non-linear logistic model of the second sandhill crane sequence was damped-stable ($1 < AQd = 1.957 < 2$). However, as it takes over 50 years to attain equilibrium following modest disturbances, the model is on the verge of instability (i.e. *AQd* is close to 2). In a variable environment, the simulated population fluctuations mimicked the typical sharp oscillations seen in the data and corresponded well with the original data in phase space (Figure 11.5).

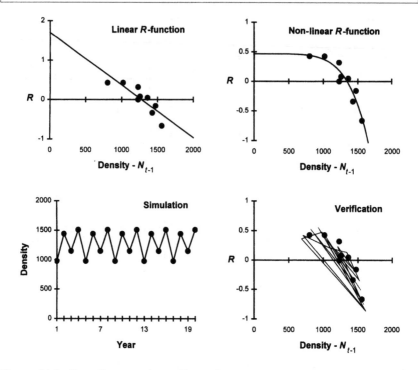

Figure 11.3 Top, linear and non-linear first-order logistic R-functions fitted to the first sandhill crane sequence (1959–1968); bottom-left, a deterministic simulation with the non-linear model; bottom-right, comparison of a stochastic simulation (thin line) with original data (dots) in phase space.

11.1.3 Analysis of the adjusted time series (1959–1978)

Analysis of the two sequences indicated that the same first-order process was operating before and after the shift in the equilibrium density. Therefore, the two sequences were spliced together after adjusting the first sequence to the mean of the second with the equation

$$N'_t = N_t + \bar{N}_2 - \bar{N}_1,$$

where N'_t is the new data series (1959–1968), N_t is the original data from the first sequence, $\bar{N}_2$ is the mean of the second sequence and $\bar{N}_1$ is the mean of the first sequence. The fit of a non-linear logistic R-function to the total adjusted series is shown in Figure 11.6 ($A = 0.477, K = 1939.66,$ $Q = 4.011, r^2 = 0.893$). Deterministic simulations showed this model to be damped-stable but on the verge of instability ($AQd = 1.913$). Stochastic simulations produced the typical saw-toothed oscillations seen in the data (Figure 11.6).

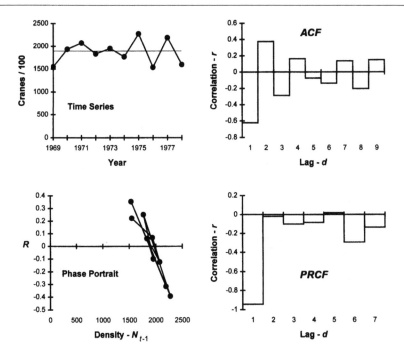

Figure 11.4   Diagnostics from the second sandhill crane sequence (1969–1978).

11.2 INTERPRETATION

The dynamics of the mid-continental sandhill crane population are apparently determined by a first-order ¯feedback process, presumably resulting from competition for non-reactive limiting resources. The curvature of the R-function ($Q > 1$) indicates that the population is regulated by *adapted intra*specific competition. This leads automatically to the hypothesis that competition for breeding territories (territorial behaviour) is the main mechanism regulating sandhill crane populations, and that the limiting factor is the number of available breeding sites.

Because all sandhill crane models are very close to the unstable condition ($AQd \geq 2$), the population should be expected to exhibit sharp oscillations even in a relatively stable environment (Figure 11.6). On the other hand, the amplitude of oscillation is limited by the curvature of the R-function and a low maximum per capita rate of change ($A \approx 0.5$). In other words, the fluctuations around equilibrium are not very large because the maximum per capita rate of change is small and this prevents large overshoots of carrying capacity. Hence, violent fluctuations in crane abundance should not be observed, even in quite variable environments.

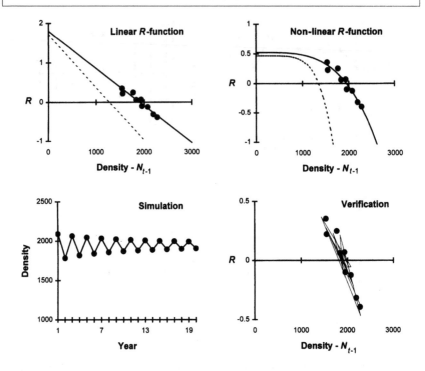

Figure 11.5 Top, linear and non-linear first-order R-functions fitted to the second sandhill crane sequence (1969–1978) (the dotted lines mark the R-functions for the first sequence); bottom-left, deterministic simulation with the non-linear model; bottom-right, verification of a stochastic simulation (thin line) against the original data (dots) in phase space.

The sandhill crane population exhibited a sudden shift in its equilibrium density between 1968 and 1969. The reason for this shift was not apparent from the data nor from information available in the literature. If the hypothesis of regulation by *intra*specific competition for breeding territories is true, then this shift could be due to an increase in the number of breeding territories available to the crane population. However, the shift could also have been caused by a change in the census procedure, hunting pressure, or in the accuracy of the sampling methods.

11.3 MANAGEMENT IMPLICATIONS

Sandhill cranes have been hunted in one or more designated areas of the United States since 1961, in Canada since 1959 and in Mexico since 1940, and the total kill has been estimated at around 18,000 cranes per year[41]. Using the non-linear logistic R-function fit to the adjusted time series

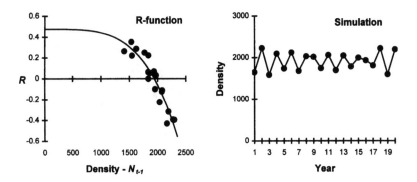

Figure 11.6 Non-linear first-order R-function fitted to the adjusted sandhill crane time series (1959–1978), and a typical stochastic simulation with standard deviation estimated from the model fit.

(Figure 11.6) to estimate the realized per capita rate of change of the crane population, R, the sustainable yield curve for the sandhill crane was calculated from the relationship $Y_n = N_h(e^R - 1)$ [see equation (10.14)], with Y_n the sustainable yield, and N_h the density of the population after harvest. This curve shows that the theoretical maximum sustainable yield is about 60,000 cranes per year in an invariant environment (Figure 11.7, left). However, when harvesting was carried out under normal random environmental disturbances, a different picture arose. Figure 11.7 (right) shows the results of running the sandhill crane model under different harvesting levels for 300 years in a randomly varying environment. In each experiment, a constant number of cranes was harvested, except when the population was below 50,000 when harvesting was suspended. The results indicate that harvesting at the theoretical maximum sustainable yield (60,000 cranes per year) is a dangerous policy. Not only is the average yield lower, but hunting would have to be curtailed in most years (250 out of 300 years) because the population was below the critical density of 50,000. The maximum sustainable yield in a variable environment is around 50,000 cranes per year with a residual (post-harvest) population of around 150,000. Hence, the current level of sandhill crane harvesting ($\approx$ 18,000 per year) appears to be sustainable provided that the environment does not deteriorate substantially.

The stochastic yield function has an interesting and significant property. Notice that the harvest rate can be increased up to 51,000 per year without significant effects on the dynamics of the crane population (Figure 11.7, right). However, even a slight increase in harvest above this level results in irregular collapses of the crane population and the suspension of harvesting in a significant number of years. In fact, harvesting at the theoretical (deterministic) level ($\approx$ 60,000 per year) results in a very low average

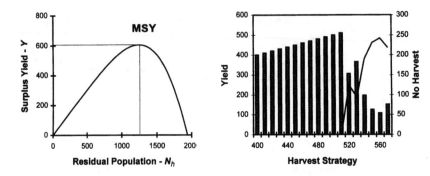

Figure 11.7 Left, sustained yield curve for the sandhill crane model showing maximum sustainable yield of $Y_t \approx 600 \times 100 \approx 60,000$ individuals with a residual (after harvest) population density $N_h \approx 125,000$. Right, results of stochastic simulations (300 years with standard deviation estimated from the data, $s = 0.082$) for different harvesting levels; the histogram shows the average yield per year and the line is the number of years when harvesting was suspended because the crane population fell below the critical density of 50,000 individuals. Remember that all figures should be multiplied by 100 to convert them to actual population numbers.

yield ($< 20,000$ per year) because hunting has to be suspended in 3 out of every 4 years. The model informs us that crane dynamics are very sensitive to harvesting near the stochastic maximum sustainable yield, and supports the conservative strategy employed by current management agencies.

11.4 EXERCISES

1. Calculate the sustained yield curve for the sandhill crane from the model of the adjusted time series. This is easy to do with a hand calculator or a spreadsheet.
2. Build a model to calculate sandhill crane yields under different harvesting strategies in a variable environment. This can also be done with the *PAS* program *P2m – Population Manager* (see Preface) if you first build a model of the adjusted series with *P1a – Single-species Time Series*, and save it to disk (data can be found in reference [41]). Compare the actual harvests with those predicted by the sustained yield curve.

12

Cone beetles: competition for food?

The annual cone mortality caused by these insects is highly variable, but is, nevertheless, somewhat predictable because their populations inexorably increase until limited by cones. (W. J. Mattson, 1986)

Cone beetles are small black insects (Order Coleoptera) belonging to the family Scolytidae, one of the most destructive families of forest insects. In this chapter the population dynamics of red pine cone beetles (*Conophthorus resinosae*) are studied over the years 1969–1978 in a red pine (*Pinus resinosae*) seed production area in Wisconsin[42].

Female red pine cone beetles fly to pine trees in late spring or early summer and bore galleries in the developing cones. Eggs are laid in short lateral galleries near the developing seed ovules. Each female may attack up to 20 or so cones. Attacked cones usually die and produce no seed. Larvae hatch from the eggs, feed on the tissues of the cone, form pupae in late summer, and emerge soon afterwards as adults. Emergent adults then bore into terminal shoots which fall to the ground. Adult beetles spend the winter on the forest floor inside the buds of terminal shoots.

Overwintering beetle populations were estimated by randomly sampling fallen shoots on the forest floor[42]. Densities of attacked and unattacked red pine cones were estimated in late summer (after beetle attacks had ceased) by counting all damaged and undamaged reproductive structures on eight randomly selected branches (stratified by height and aspect) on 10 randomly selected pine trees.

12.1 ANALYSIS OF CONE BEETLE DYNAMICS

The cone beetle population data and diagnostics are shown in Figure 12.1. Although there is some indication of a trend in the time series (shown

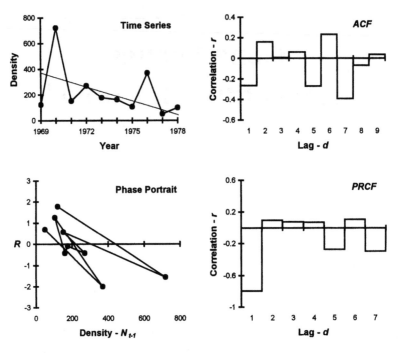

Figure 12.1 Time series data and diagnostics for red pine cone beetles (numbers per 0.1 acre) in a Wisconsin seed production area[42].

by the sloping line through the series), the *ACF* and return time statistics indicate a stationary mean (*MRT* = 0.842; *VRT* = 0.668). Analysis of detrended and original time series produced similar models, so the original data were used. This is purely a matter of personal preference, however, and others may wish to model the trend.

The sharp saw-toothed oscillations of the time series, narrow phase portrait, and dominant *PRCF* at lag 1 indicate that the dynamics were governed by a first-order ¬feedback process. Hence, the data were fitted to a first-order, linear logistic *R*-function with parameters A = 1.069, K = 231.858, Q = 1, d = 1, r^2 = 0.572 (Figure 12.2). The fit was improved by using a non-linear model with parameters A = 4.071, K = 30.819, Q = 0.308, d = 1, r^2 = 0.650 (Figure 12.2). The use of Gompertz or two-lag models did not improve the fit.

Deterministic simulations showed the non-linear logistic cone beetle model to be damped-stable in a constant environment. This is, of course, obvious from the stability criterion AQd = 1.25, which is in the damped-stable domain (e.g. 1 < AQd < 2, see Chapter 7). Stochastic simulations mimicked the typical sharp oscillations seen in the data (Figure 12.2). The phase portrait of simulated and real data showed good correspondence, verifying that the model behaved in a manner consistent with the real system (Figure 12.2).

12.2 INTERPRETATION

The dynamics of red pine cone beetle populations are apparently determined by a first-order ⁻feedback mechanism, probably due to *intra*specific competition for a passive limiting resource. The most reasonable hypothesis is that cone beetle production is dependent on the availability of food in the form of red pine cones.

Analysis of the model suggests that cone beetle numbers are very stable in a constant environment. Hence, fluctuations in numbers must be largely due to external environmental factors. If the hypothesis of regulation by *intra*specific competition for cones is correct, then fluctuations in beetle numbers should be largely due to year-to-year variability in cone production, and the per capita rate of change of the beetle should be positively related to the abundance of pine cones, or inversely related to

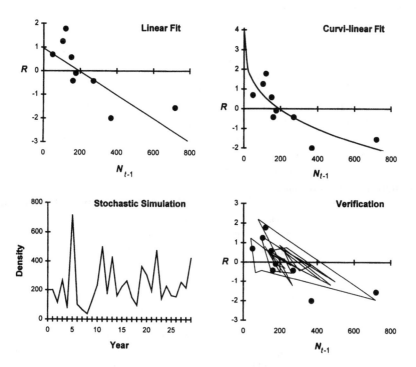

Figure 12.2 Linear and non-linear first-order logistic R-functions fitted to the cone beetle data with stochastic simulation by the non-linear model (line) verified against the original data (dots).

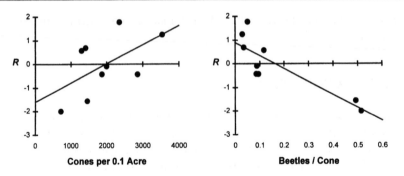

Figure 12.3　Linear regression analysis of the cone beetle per capita rate of change on the density of cones available in a given year (left) and the beetle/cone ratio (right) ($r^2 = 0.316$ and 0.722, respectively).

the beetle/cone ratio [i.e. as in equation (6.4)]. These expectations are supported, in general, by the data (Figure 12.3).

　　Support for the hypothesis can also be obtained by fitting the data to the two-species logistic consumer–resource model. The equation for a population of consumers is [equation (6.4)]

$$R_P = A_P - C_P P_{t-1} - C_{PN} \frac{P_{t-1}}{N_{t-1} + F},$$
(12.1)

with R_P the estimated per capita rate of change of the beetle population, P the density of beetles and N the density of cones. The following parameter estimates were obtained by multiple linear regression: $A_P = 0.948$, $C_P = 0.0006$, $C_{PN} = 4.932$, $F = 0$ (this beetle only feeds on red pine cones), with a coefficient of multiple determination $rm^2 = 0.725$. Notice that this model provides a better fit to the data than the non-linear logistic (i.e. $rm^2 = 0.725$ versus $r^2 = 0.572$). In fact, the simple coefficient of determination between R_P and the consumer/resource ratio (P/N) is 0.722, which means that the relationship with the food supply provides most of the explanatory power of the model (Figure 12.3, left).

　　The hypothesis that pine cones act as a non-reactive resource can be tested by fitting the data to the model for a resource population [equation (6.5)]

$$R_N = A_N - C_N N_{t-1} - C_{NP} \frac{P_{t-1}}{N_{t-1} + F},$$
(12.2)

where R_N is the estimated per capita rate of change of the cone population. Multiple regression analysis gave the following parameter estimates: $A_N = 2.118$, $C_N = 0.000097$, $C_{NP} = 1.454$, $F = 0$, with a coefficient of multiple determination $rm^2 = 0.862$. In this case, however, it was the simple coefficient of determination between R_N and the density of cones,

N, that explained most of the variability in cone rates of change ($r^2 = 0.778$), while the effect of the beetle was insignificant ($r^2 = 0.09$). The conclusion that beetle densities have no effect on the abundance of pine cones supports the hypothesis that cone beetle numbers are regulated by *intra*specific competition for a non-reactive resource (pine cones), and that fluctuations in beetle density are a result of variability in the numbers of pine cones produced each year.

12.3 MANAGEMENT IMPLICATIONS

Red pine cone beetles destroy up to 100% of the cone crop in the studied seed production area[42] and, therefore, management of this pest is essential if a seed crop is to be harvested. One of the things the manager needs to know is whether controlling the beetle population can reduce seed losses. Of course, this strategy would only be effective if damage to the cone crop was related in one way or another to the size of the beetle population. However, regression analysis of the number of cones attacked per year against beetle density at the beginning of the year showed no significant relationship (Figure 12.4, left). On the other hand, there was a strong relationship between the number of cones attacked and the size of the cone crop that year (Figure 12.4, right). These results imply that cone damage is independent of beetle population density in the seed production area and, therefore, that the control of the beetle population in this area would be pointless. The probable reason for this lack of correspondence between beetle numbers and crop losses is that large cone crops attract beetles into the area from the surrounding forest. In fact, it seems likely that the beetle "population" being studied is really a *local* population, in the sense of the definitions in Chapter 2. The lesson for

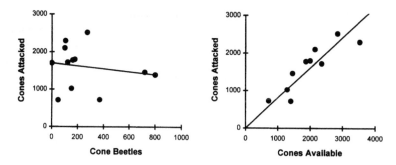

Figure 12.4 Linear regression analysis of the number of cones attacked on the density of cone beetles present in the seed production area (left) and the initial density of the cone crop (right) ($r^2 = 0.015$ and 0.775, respectively).

the pest manager is simple: damage to cone crops can only be reduced by controlling beetles entering the seed production area (e.g. by using traps), or by protecting the cones from attack (e.g. by treating the cones with insecticides or repellents). In fact, one of the most successful practises in cone beetle control is to inject systemic insecticides into the tree. These pesticides are carried in the sap stream and concentrate in growing tissues, like the cones, thereby killing the beetles before they can penetrate into the cone.

This analysis illustrates a common problem of applying population dynamics models to data that represent only a fraction of the real population. Although the conclusions – that cone beetle dynamics are governed by *intra*specific competition for food, and that fluctuations in beetle numbers are a reflection of year-to-year variation in cone abundance – are probably true, the models only represent the dynamics of a local population and, therefore, are of little use for forecasting crop losses or developing pest management strategies.

12.4 EXERCISES

1. Build a model of the interaction between red pine cones and cone beetles by fitting equations (12.1) and (12.2) to the field data[42] for cone beetles and red pine cones, respectively. This can be done with standard statistical programs that have the capability to perform multiple regression analysis, or with the *PAS Two-species Time Series* program *P2a* (see Preface for details). Note that if you use *P2a*, you will first have to run the data for both species through the *PAS Single-species Time Series* program *P1a* because the single-species model is the default in the case of insignificant, or nonsensical, interactions between the two species. If you use *P2a*, remember to save the best model to disk.

2. Use the parameters estimated above to calculate and plot the equilibrium isoclines for beetles and cones. Once a two-species model has been built, these calculations can be done with the *PAS* program *P2b* – *Two-species Modelling Simulation*, or on a spreadsheet by using equations (6.6) and (6.7) to calculate the isoclines for beetles and cones, respectively.

3. Simulate the dynamics of the two interacting species in constant and noisy environments and plot the trajectories in the time domain and phase space. These calculations can be done on a spreadsheet or with the *PAS* program *P2b*.

Spruce needleminers: parasitoids?

13

Parasitism is obviously the most important factor in limiting and regulating the pest insect. (M. Munster-Swendsen, 1985)

The spruce needleminer (*Epinotia tedella*) is a small moth that lives in European spruce forests. It belongs to the family Tortricidae, one of the most destructive families of the Order Lepidoptera (moths and butterflies). Needleminer moths emerge from the forest floor in June and females lay eggs singly on spruce needles. Upon hatching, the larvae bore into the needle and feed on the soft tissues within. In November the mature larvae spin down to the forest floor where they hibernate in silken cocoons. Larvae pupate within their cocoons the following May and emerge as adults soon afterwards. Needleminer outbreaks may occasionally defoliate spruce forests but are typically of short duration (1–2 years) so that trees are able to recover quickly and control is rarely needed.

Young spruce needleminer larvae are attacked by a number of parasitic wasps (Order Hymenoptera), but only two of these, *Pimplopterus dubius* (Family Ichneumonidae) and *Apanteles tedellae* (Family Braconidae) are major primary parasitoids. As we shall see, parasitic wasps seem to play an important role in the population dynamics of the spruce needleminer.

The data used in this analysis were obtained during a 19-year study of spruce needleminer populations in the forests of North Zealand, Denmark[43]. Needleminer populations were sampled by intercepting descending larvae with 200 funnel traps placed under trees from November to December. Larvae captured in the traps were dissected to determine if they were parasitized.

13.1 ANALYSIS OF NEEDLEMINER DYNAMICS

Spruce needleminer population data and diagnostics are shown in Figure 13.1. The *ACF* and return time statistics indicate a stationary series (MRT = 2.44; VRT = 2.29) with a cycle period of 5–8 years. The cyclical oscillations, broad circular phase portrait, and dominant *PRCF* at lag 2 indicate that the dynamics were dominated by a second-order (delayed) feedback process, as expected from the action of the fourth principle. Fitting a second-order, non-linear logistic *R*-function to the data yielded the parameters A = 2.122, K = 818.66, Q = 0.365, d = 2, and coefficient of determination r^2 = 0.653 (Figure 13.2). The use of Gompertz or two-lag models did not improve the fit.

Deterministic simulations demonstrated that the spruce needleminer model was damped-stable (Figure 13.2). The stability criterion, AQd = 1.549, also indicates damped-stable steady-state behaviour. Notice that the deterministic simulations have a period of 6–7 years.

Stochastic simulations were characterized by cyclic oscillations with variable amplitude and 5–7-year period, much like the original data, and simulated trajectories showed good correspondence to the real data in phase portrait verification (Figure 13.2, bottom).

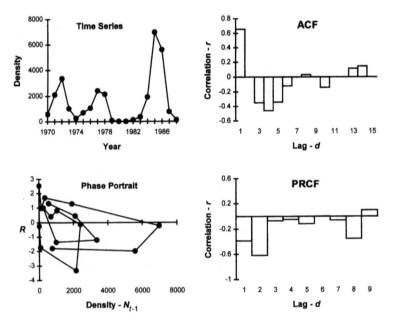

Figure 13.1 Time series data and diagnostics for spruce needleminer larvae descending to the forest floor in November–December in a spruce stand in Denmark (numbers per 10 square metres of land area)[43].

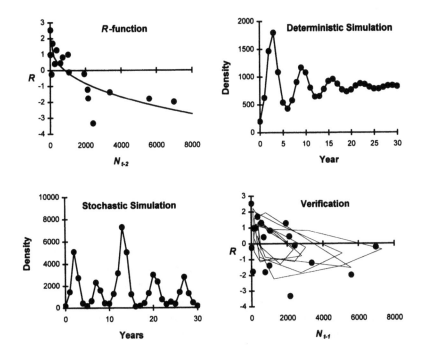

Figure 13.2 Non-linear second-order logistic R-function fitted to the spruce needleminer data with deterministic (top) and stochastic (bottom) simulations, and verification of simulated (thin line) and original data (dots) on the phase portrait.

13.2 INTERPRETATION

The dynamics of the spruce needleminer appear to be dominated by second-order ⁻feedback, indicating an interaction with reactive environmental factors (the fourth principle). One plausible hypothesis is that the guild of insect parasitoids attacking the larval stage is responsible for the second-order dynamics. This hypothesis can be tested because data were collected on the number of larvae parasitized each year.

If the hypothesis is true then there should be strong feedback between needleminer and parasitoid populations, in which case the per capita rate of change of the needleminers (R_N) and parasitoids (R_P) should have a strong inverse relationship with the parasitoid/needleminer ratio, in accordance with equations (6.5) and (6.4), respectively. Regression analysis indicated that over 70% of the variation in needleminer and parasitoid rates of change can be explained by the interaction between the two populations, and supports the hypothesis that the second-order dynamics are the result of this interaction (Figure 13.3).

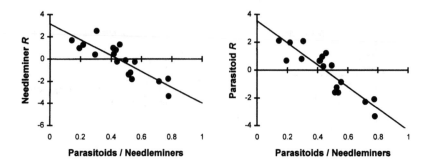

Figure 13.3 Left, regression analysis of the per capita rate of change of the spruce needleminer against the parasitoid/needleminer ratio ($r^2 = 0.726$). Right, regression analysis of the parasitoid per capita rate of change against the parasitoid/needleminer ratio ($r^2 = 0.795$).

Because of the large coefficients of determination in the feedback between needleminer and parasitoid populations, it seems unlikely that other reactive environmental factors can contribute very much to the cyclic dynamics. In fact, an analysis of a detailed simulation model, which included interactions with food, disease, secondary parasitoids (hyper- and clepto-parasitoids) and weather, also concluded that parasitoids were the main factor driving the population cycles[43].

13.3 MANAGEMENT IMPLICATIONS

Because the dynamics of spruce needleminer populations seem to be strongly influenced by parasitoids, it is important for forest managers to consider the needleminer–parasitoid interaction if they are to understand the population dynamics of the pest, and if they want to forecast and manage pest numbers. One way to do this is to study models of the needleminer–parasitoid interaction. For example, models can be built by fitting the data to two-species R-functions such as equations (6.4) and (6.5); i.e.

$$R_P = A_P - C_P P_{t-1} - C_{PN} \frac{P_{t-1}}{N_{t-1}}, \text{ parasitoid } R\text{-function} \qquad (13.1a)$$

$$R_N = A_N - C_N N_{t-1} - C_{NP} \frac{P_{t-1}}{N_{t-1}}, \text{ needleminer } R\text{-function} \qquad (13.1b)$$

The following parameter estimates were obtained by multiple regression: for the parasitoid R-function, $A_P = 3.5$, $C_P = 0.000075$, $C_{PN} = 7.685$, with a coefficient of multiple determination $rm^2 = 0.797$, and for the needleminer

R-function, $A_N = 3.2$, $C_N = 0.000105$, $C_{NP} = 6.819$, and $rm^2 = 0.743$. Steady-state analysis showed that the interactive model was damped-stable in a constant environment (Figure 13.4, left). In a variable environment, however, the model exhibited persistent oscillations similar to those observed in the data (Figure 13.4, right).

13.3.1 Biological control

Biological control, or biocontrol, is the use of natural enemies to reduce and maintain pest populations at densities where damage to the crop is inconsequential. A manager may be interested in knowing, for instance, if parasitoids are *effective* biocontrol agents against the spruce needleminer.

The effectiveness of a biocontrol agent depends on two basic properties of the interaction between natural enemy and pest populations:

1. *Suppression.* The first measure of effectiveness is the degree to which the biocontrol agent suppresses the pest population below its potential level. Suppression, therefore, measures the difference between the pest equilibrium in the presence of parasitoids (J) and the carrying capacity of the pest in the absence of parasitoids (K). The degree of suppression is usually expressed as a percentage $S = 100 (K - J)/K$. The equilibrium density of the needleminer in the presence of parasitoids was calculated, from equations (13.1), to be $J = 1203$ larvae per 10 square metres, while its carrying capacity in the absence of parasitoids was $K = 30,359$. Hence, the degree of suppression of needleminer densities by parasitoids is $S = 100 \times (30,359 - 1203)/30,359 = 96\%$. In other words, parasitoids are very effective at suppressing spruce needleminer populations below their potential levels of abundance.

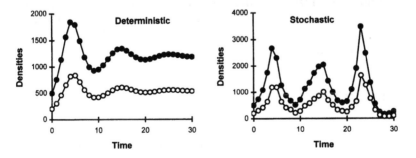

Figure 13.4 Simulation of spruce needleminer and parasitoid population dynamics using equation (13.1) and the stochastic forecasting equation (10.13) in constant (standard deviations $s_N = s_P = 0$; left) and variable environments ($s_N = 0.2$; $s_P = 0$; right).

2. *Stability*. The second measure of effectiveness is the stability of the community equilibrium created by the interaction with the biocontrol agent. As a general rule, the greater the degree of suppression by a biocontrol agent, the lower the stability of the interaction, all else being equal. In the case of the needleminer–parasitoid model, the equilibrium point is known to be stable but the population can exhibit cyclical fluctuations of considerable amplitude in a variable environment (Figures 13.2 and 13.4). Because of this, occasional outbreaks of the needleminer are to be expected. When outbreaks occur, the forest manager may be tempted to take action to prevent damage to the forest. However, our analysis indicates that pest control operations should carefully consider the effects on the parasitoids. For instance, spraying with a pesticide that kills the parasitoids will usually be followed by a rebound of the pest population to much higher densities. Even spraying with a selective pesticide that only kills the needleminer can result in pest rebound as the parasitoids collapse from lack of food.

13.3.2 Forecasting

Given estimates of the densities of needleminers in the current year and the number parasitized, the expected density next year can be forecast from the needleminer R-function [equation (13.1b)] and the familiar step-ahead forecasting equation (10.1). However, knowing that there is a strong relationship between parasitism and needleminer rates of change, it should be possible to design a simpler forecasting procedure based on the proportion of the needleminer population parasitized. For example, Figure 13.3 shows how needleminer populations tend to increase ($R > 0$) when less than 0.45 of the larvae are parasitized and to decrease ($R < 0$) when parasitism is greater than 0.45 (note that the P/N ratio is the same as the proportion parasitized). Hence, by simply observing the proportion of the needleminer population parasitized in a given year, the manager can make a qualitative forecast of the direction of needleminer population change in the following year.

13.4 EXERCISES

1. Build a model of the interaction between spruce needleminer and parasitoid populations by fitting equations (6.4) and (6.5) to the field data[43]. This can be done with standard statistical programs that have the capability to perform multiple regression analysis, or with the *PAS Two-species Time Series* program *P2a* (see Preface for details). Note that if you use *P2a*, you will first have to run the data for both species

through the *PAS Single-species Time Series* program *P1a* because the single-species model is the default in the case of insignificant, or nonsensical, interactions between the two species. If you use *P2a*, remember to save the best model to disk.

2. Use equations (6.6) and (6.7) to calculate and plot the equilibrium isoclines for needleminers and parasitoids. Once a two-species model has been built, these calculations can be done on a spreadsheet or with the *PAS* program *P2b – Two-species Modelling Simulation.*

3. Simulate the dynamics of the two interacting species in constant and noisy environments and plot the trajectories in the time domain and phase space. These calculations can be done on a spreadsheet or with the *PAS* program *P2b.*

4. Use the model of the interaction between spruce needleminers and their parasitoids and the step-ahead forecasting equation to predict the number of needleminers expected in 1989 from the numbers present in 1988.

14

Mountain pine beetles: co-operation?

When it is present in a forest in epidemic numbers it shows no discrimination as to the apparent health and vigor of the trees attacked – at such times, indeed, seeming to show a slight preference for trees of more than average vigor. When not epidemic, however, it breeds in trees weakened by mistletoe, lightning, or other causes. (M. W. Blackman (1931) writing about the Black Hills beetle, later recognized as the same insect as the mountain pine beetle.)

In the preceding three chapters the diagnostic procedures developed in Chapter 9 were used to analyse, interpret and model a continuous series of observations of real populations changing through time (a time series). This approach has an appealing objectivity because it requires no preconceived notions about the actual mechanisms involved in the observed dynamics. However, situations often arise where time series data are lacking or are insufficient to perform this kind of analysis. In this chapter, I show how an understanding of the causes of population dynamics can be developed from fragmentary time series data and knowledge of the biology, behaviour and ecology of the population system.

Mountain pine beetles (*Dendroctonus ponderosae* Hopkins)[44] are small brown beetles (Coleoptera: Scolytidae) that inhabit the pine forests of western North America. The Family Scolytidae is one of the most destructive groups of forest insects, and the mountain pine beetle is perhaps the most dangerous of all. Adult mountain pine beetles fly in mid-summer and bore galleries under the bark of living pine trees. Eggs laid in these galleries hatch into larvae that then feed on the inner bark tissues. The winter is usually spent in the larval stages and pupation generally occurs in the following spring or summer. Shortly after this, the adult beetles emerge from the dead tree and fly to attack new hosts.

Successful breeding of bark beetles is contingent on the death of all or part of the tree and, to accomplish this end, attacking beetles inoculate *pathogenic fungi* as they bore into the tissues of their living host. The fungi grow through the tissues, killing the cells and making the environment suitable for the growth of beetle larvae. While adult beetles are boring into the bark they also produce chemicals called *aggregating pheromones* that attract nearby beetles to the tree under attack. If the surrounding beetle population is large enough, the resulting *mass attack* may induce sufficient fungal infections to overcome the resistance of the tree, in which case the tree dies and a new generation of beetles is born. If the mass attack is insufficient and the tree is reasonably healthy, the tree can secrete sufficient defensive chemicals, mostly terpenoid compounds, to "pitch out" the beetles and the tree will usually survive. Hence, whether a particular tree lives or dies depends on its innate resistance to attack and the number of beetles flying in the immediate vicinity. Because of this, sparse beetle populations are unable to colonize healthy trees and are, therefore, restricted to recently killed or dying trees, such as those struck by lightning, blown down by wind, or infected by other insects or diseases. When beetle populations are large, however, healthy trees can be killed by rapid and massive attack.

Mountain pine beetles are used as food by a diverse group of organisms, including woodpeckers, insect parasitoids and predators, and micro-organisms, but none seem to have much impact on the dynamics of the beetle population[44]. In other words, the *limiting* factor for mountain pine beetle populations seems to be the quantity of susceptible host trees and, because of the "mass attack" phenomenon, this quantity is directly related to the size of the beetle population.

14.1 INTERPRETATION OF PINE BEETLE DYNAMICS

Although the population dynamics of the mountain pine beetle have been studied intensively for more than 50 years, there are no continuous time series that cover the complete range of population dynamics (although, as we shall see below, there are fragmentary data series). Hence, the interpretation of mountain pine beetle dynamics must rest, in large part, on a general qualitative understanding of mountain pine beetle biology and ecology and, in particular, its attack behaviour in relation to pine tree resistance.

Mountain pine beetle populations persist in pine forests at sparse densities for long periods of time by infesting those trees weakened by diseases, insects, weather, overcrowding, or old age (Figure 14.1). Beetles that bore into normal healthy trees cannot kill them because there are not enough

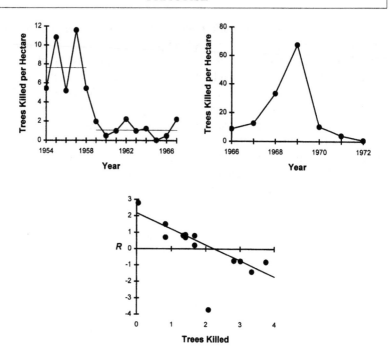

Figure 14.1 Left, population dynamics of the mountain pine beetle under endemic conditions (data from Starvation Ridge, Glacier National Park). Right, population dynamics of the mountain pine beetle under epidemic conditions (data from Yellowstone National Park). The data report the number of pine trees killed each year per hectare of forest[44]. Bottom, a linear logistic R-function fitted to the Starvation Ridge data after adjustment to a common mean; $A = 2.18$, $C = 0.97$, $r^2 = 0.89$.

"backup" beetles to overwhelm their defences. Beetle populations may build up in restricted localities, or *epicentres*, but attacks on healthy trees in adjacent stands are rejected so that the population does not spread out from the epicentre. Under these conditions the beetle population is considered by forest managers to be in the *endemic* state.

On occasion, local beetle populations may increase for one reason or another and, if numbers are sufficient to overcome the resistance of normal healthy trees, the population may spread into adjacent stands. Alternatively, widespread stress on the forest, as might result from several years of severe drought or excessive rain (root flooding), may lower the resistance of large areas of forest to the point where local populations can expand out of their epicentres. As the population grows, more and more trees can be killed and the population can then sweep over vast regions (Figure 14.2). The beetle population is then considered to be in the *epidemic* state.

14.2 DEDUCTION OF THE PINE BEETLE *R*-FUNCTION

In Chapter 10 the following procedure was suggested for deducing the qualitative form of the *R*-function for a particular population:

1. Determine the limiting factor or factors.
2. Determine if there is the potential to escape from the constraints imposed by the limiting factor(s) and, thereby, whether one or two *R*-functions need to be considered.
3. Determine for each *R*-function if the limiting factor is reactive or non-reactive and, from this, if time delays are likely to be encountered in the ‾feedback process.
4. Determine if non-linearities in the *R*-function should be considered.
5. Attempt to obtain estimates of the parameters of the *R*-function.

The empirical evidence indicates that the availability of weakened or dying pine trees is the main factor limiting the abundance of the mountain pine beetle under normal endemic conditions. However, the population can *escape* from these limits when numbers reach the critical density required to overwhelm the resistance of healthy trees. When this occurs, the number of susceptible trees (limiting factor) can increase by several orders of magnitude (Figure 14.1).

It appears that two *R*-functions will be required to describe the dynamics of mountain pine beetle populations, one for limitation by weakened or dying hosts when beetle populations are sparse and another for limitation by the total host population, including weakened and healthy trees, when beetle populations are high. There must also be an unstable escape threshold that separates the domains of attraction to these two *R*-functions, and this threshold must be directly related to the overall health or vigour of the forest, for more beetles are required to overcome more resistant hosts. Thus, vigorous forests should have higher escape points, with very young or exceptionally vigorous forests perhaps being immune to attack.

The next thing to consider is whether the limiting factors are likely to be reactive or non-reactive. When beetle populations are low, the number of susceptible trees available each year is determined by exogenous environmental factors that lower the resistance of individual pines; for example, storms, disease, and infestations by other insects. Therefore, the annual food supply is independent of beetle numbers and acts as a non-reactive factor. Under these conditions, the beetle population will be regulated by competition for food (third principle) and the dynamics should be characterized by first-order saw-toothed oscillations (e.g. Figure 14.1, top-left). On the other hand, when beetle populations are high they can destroy most of the trees in the forest and the limiting factor becomes reactive; i.e. the quantity of food available in a given year is now dependent

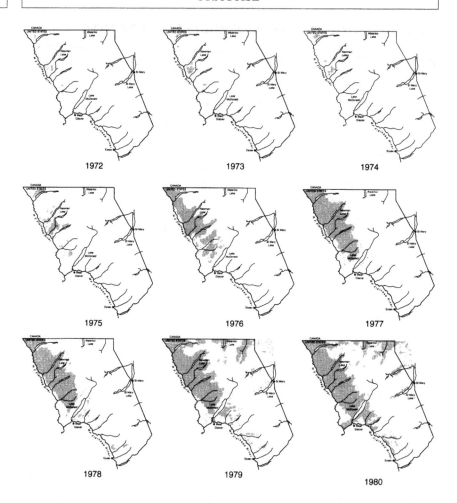

Figure 14.2   Spread of a mountain pine beetle outbreak through Glacier National Park, Montana[44].

on the density of the beetle population in previous years. Under these conditions the fourth principle should be evoked and the dynamics characterized by higher-order cycles (e.g. Figure 14.1, top-right).

Fitting R-functions to data from the endemic population (Figure 14.1, left), following adjustment to a common mean density (Chapter 10), suggests that the rate of tree killing can be described by a linear, first-order, logistic model (Figure 14.1, bottom). The endemic equilibrium density, J, can be estimated from the relationship, $J = A/C = 2.18/0.97 = 2.25$ trees per hectare. Unfortunately, there are not enough data to estimate the equilibrium density of epidemic populations so indirect methods must be used. What we need is an estimate of the long-term sustainable equilibrium density when the total forest is susceptible to attack. The beetle population can only be sustained indefinitely if it kills no more than the annual production of the forest, or the number of tree equivalents that are produced annually by growth and reproduction. An estimate of this quantity was obtained by consulting yield tables for thinned 80-year-old lodgepole pine stands[44]: stand density = 900 trees per hectare, total volume = 200 cubic metres per hectare, and mean annual incremental growth = 2.7 cubic metres per hectare. From these figures it is possible to calculate the volume of an average tree; i.e. 200/900 = 0.22 cubic metres. Hence, the annual production in tree equivalents is 2.7/0.22 = 12.3 trees per hectare per year. These 12.3 tree equivalents can be utilized each year without depleting the stand and so K = 12.3 trees per hectare. Because the tree population acts as a reactive limiting factor during epidemics, the dynamics can probably be described by a second-order (d = 2) linear logistic R-function with parameters $A_J = A_K$ = 2.18 and C_K = 2.18/12.3 = 0.18.

Finally an estimate of the escape threshold, E, is required. As mentioned earlier, this quantity will vary with the vigour of the stand so we can assume any value or, if we wish, let the value vary with time or stand age. In order to create interesting simulations, however, the threshold should be placed somewhere between J and K, say E = 5 infested trees per hectare. The two R-functions and separatrix are shown in Figure 14.3.

14.3 SIMULATED POPULATION DYNAMICS

It is now possible to simulate the dynamics of mountain pine beetle populations in variable environments using the model

$$R = \begin{cases} 2.18 - 0.18N_{t-2}; \ N_{t-1} > 5 \ldots \text{epidemic} \\ 2.18 - 0.97N_{t-1}; \ N_{t-1} \leq 5 \ldots \text{endemic} \end{cases} \quad (14.1)$$

An example is shown in Figure 14.4. The endemic period (years 0–24) and epidemic period (years 24–30) are plotted separately to make comparison with the data in Figure 14.1 easier. Notice how the endemic population fluctuates around its low-density equilibrium with sharp saw-toothed oscillations, similar to the data from Starvation Ridge (Figure 14.1, left), while the outbreak dynamics are similar to the Yellowstone data (Figure 14.1, right). A pleasant surprise is that the duration of the outbreak and the rate of tree mortality are very similar in both simulated and natural outbreaks. Of course, we could also simulate situations in which the stand aged gradually leading to lower resistance and lower escape thresholds as time progressed.

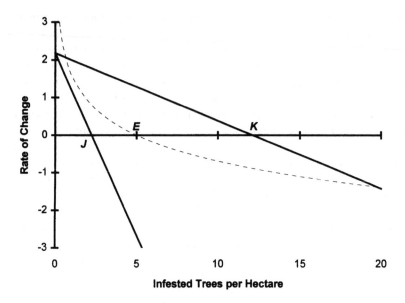

Figure 14.3 Two *R*-functions for the mountain pine beetle separated by an unstable separatrix (broken line) with parameters estimated as described in the text.

It is important to realize that the spatial dimension of a mountain pine beetle population changes drastically during outbreaks. The model of the endemic population may represent the true population (see Chapter 2 for the definition of population), but during outbreaks the model really describes a *local* beetle population. However, the dynamics can be simulated over large areas by describing the mosaic of stands of different sizes and ages over an extensive landscape, and then simulating the dynamics of tree mortality using equation (14.1) plus a dispersal process. For example, beetles could be allowed to invade adjacent stands when the density of killed trees in the epicentre reaches the escape threshold of these stands. Spatially explicit models of this type predict that mountain pine beetle outbreaks, once initiated, may spread over large areas. In other words, this bark beetle exhibits *eruptive* outbreaks (Chapter 8).

14.4 MANAGEMENT IMPLICATIONS

The mountain pine beetle model predicts that beetle populations will remain at low densities (around 2 trees per hectare per year) for long

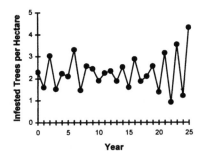
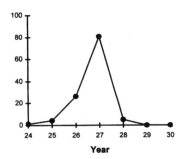

Figure 14.4 Dynamics of a simulated mountain pine beetle population in a noisy environment using the model shown in Figure 14.3 and equation (14.1); the long endemic period, years 0–25 (left) and short epidemic period, years 24–30 (right).

periods of time, but will occasionally erupt to extremely high densities (killing up to 80 trees per hectare per year). Outbreaks are likely to be precipitated by environmental disturbances that affect forest health and lower the escape threshold. Furthermore, because emigration from outbreak areas can raise beetle densities in adjacent stands above the escape threshold, the model predicts that outbreaks will spread over vast landscapes containing pine stands of varying susceptibility. The lessons for the forest manager are fairly obvious:

1. Practise forest sanitation by harvesting old, diseased and/or damaged trees. This will remove the food supply for the endemic population and thereby maintain it below the escape threshold.
2. Improve forest vigour by thinning overstocked stands to reduce stress from competition, and fertilizing nutrient-deficient soils. This will increase the escape threshold.
3. Control beetle populations at the epicentres by removing infested and/or susceptible trees; i.e. reduce beetle populations so that they remain below the escape threshold.
4. Vigorous control efforts are required during periods when the forest is under stress (e.g. during droughts or floods) to prevent bark beetle populations from escaping and spreading into healthy forests. It is important to realize that the logistics of control become more difficult, or even impossible, after beetle populations have spread over large areas. In other words, if beetle populations increase above the escape threshold, for whatever reason, there is not much time before they get beyond the manager's control. Drastic measures, including clear-cut logging, controlled burns, and mass trapping with pheromone-baited traps, may be justifiable at these times.

14.5 EXERCISES

Use the model for the mountain pine beetle (14.1) to test various strategies to reduce damage from this important forest insect*:

1. As a control, run the model with parameters given in equation (14.1) for several 100-year periods and different levels of random variability to get a feel for the kinds of dynamics and outbreak frequencies that can be expected without management actions.

2. Simulate a strategy of improving forest hygiene (removing susceptible hosts) by reducing the endemic carrying capacity to 1 susceptible tree per hectare per year. Examine the effect of this strategy on the dynamics and frequency of beetle outbreaks.

3. Simulate a strategy of improving stand vigour and resistance to attack (thinning the stands) by increasing the escape threshold to 8. Examine the effect of this strategy on the dynamics and frequency of beetle outbreaks.

4. Simulate a strategy of controlling the beetle population (removing infested trees or pheromone trapping) by removing 50% of the infested trees each year. Examine the effect of this strategy on the dynamics and frequency of beetle outbreaks.

*Simulations can be done on a spreadsheet or with *P1b*, the *Single-species Modelling Simulation* program in *PAS*.

Rock lobsters: predation?　　15

Rock lobsters transferred to Marcus Island were overwhelmed and consumed by whelks, reversing the normal predator–prey relation between the two species. (A. Barkai and C. McQuaid, 1988)

Until now, all the analyses in this book have been based, to some degree, on observations of population fluctuations through time (i.e. time series data). There are occasions, however, when no such data are available and so purely qualitative approaches are necessary. In this chapter we examine such an example.

Off the west coast of South Africa are two nearby islands, Malgas and Marcus Islands, that have very different benthic (ocean bottom) communities[45]. Both islands lie within a rock lobster (*Jasus lalandii*) reserve, but although lobsters are abundant at Malgas Island, they are absent at Marcus Island only 4 kilometres away. At the latter location, the benthic community is dominated by beds of mussels and large populations of sea cucumbers, sea urchins, and whelks (*Burnupena* sp.). At Malgas Island, on the other hand, communities are dominated by high densities of rock lobsters and seaweeds, but a relative scarcity of the other organisms.

Rock lobsters are an important commercial fishery in South Africa and it seems strange at first that they are absent from Marcus Island. Local fishermen report that lobster populations were originally similar at both islands. In addition, there are no discernable differences in ocean temperature, turbidity or nutrients, and rock lobsters confined in cages survive equally well at both places. However, when lobsters are freed into the benthic community at Marcus Island, they are quickly killed by the mass attack of hundreds of whelks. Apparently it is the high densities of whelks around Marcus Island that prevent the re-establishment of lobsters.

Rock lobsters are active predators and scavengers that, although preferring to feed on mussels, will eat any animal matter, including whelks. In areas where lobsters are abundant, they appear to operate as "keystone predators"[25], in that they regulate their prey species, including mussels and whelks, at relatively sparse densities and this permits the establishment of other species such as seaweeds. Under these conditions, whelks cannot become numerous enough to mass attack and kill lobsters. On occasion, however, extensive mortality of lobsters has been observed following sharp declines in sea water oxygen levels. In the event of such a disturbance, whelk populations may be able to increase to such an extent that they can prevent the re-establishment of rock lobsters. In fact, a period of low oxygen was observed in the waters off Marcus Island in the early 1970s.

15.1 QUALITATIVE ANALYSIS

Apparently the benthic community off the coast of South Africa can exist at two stable states:

1. A community equilibrium with a high density of rock lobsters ($N_m^* \cong$ 10 lobsters per square metre) and a low density of whelks ($P_m^* \cong 10$ per square metre)[45]. This community equilibrium is identified by $M = N_m^*, P_m^*) \cong (10, 10)$ in Figure 15.1. The symbols N^* and P^* identify points on the lobster and whelk zero-growth isoclines, respectively (see Section 6.3.3), while N_m^* and P_m^* are the particular points where the two isoclines cross at the community equilibrium, M.
2. An equilibrium on the whelk isocline where whelks are very abundant ($P_k^* \cong 300$ per square metre), but where there are virtually no lobsters ($N_k = 0$ per square metre) (Figure 15.1). This coordinate is located on the whelk axis, the carrying capacity for whelks in the absence of lobsters, $K = (N_k, P_k^*) \cong (0, 300)$.

Having located the approximate position of the known equilibrium points, it is now possible to search for plausible isoclines for the two species. First consider the whelk isocline, P^*, which has to pass through the points K and M. The simplest structure that satisfies these conditions is a straight line through these two points (the broken line P^* in Figure 15.2). Now, as the lobster isocline must also pass through the community equilibrium, M, let us represent it for the moment by a vertical line through M (the solid line N^* in Figure 15.2). Notice that the isoclines must cross in this way to create a stable equilibrium at M.

Now if the two equilibrium points, K and M, are truly alternative stable states of the same system, then there has to be an unstable equilibrium point between them, $U = N_u^*, P_u^*$. Notice that both isoclines must also pass through this unstable point. The simplest way to meet these conditions (perhaps the only way) is for the lobster isocline to bend downwards so that it intersects the whelk isocline at the unstable point U (Figure 15.3).

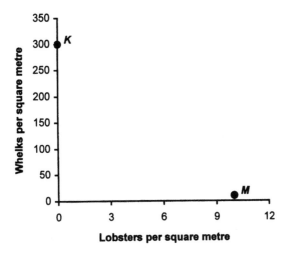

Figure 15.1 The two equilibrium points in lobster–whelk phase space; K, the whelk carrying capacity in the absence of lobsters and M, the community equilibrium where both lobsters and whelks coexist.

Of course, the exact shape of the lobster isocline, and the exact location of the unstable equilibrium point, are unknown. However, it may be possible to make an educated guess at the point where the lobster isocline intercepts the whelk axis, (N_0^*, P). This point represents the maximum density of whelks within which a single lobster can survive; i.e. if the density of whelks is exactly this number, a single lobster can survive, but if it is slightly higher the lobster will be killed. A reasonable estimate is $(N_0^*, P) \cong (0, 50)$ (Figure 15.3).

15.2 R-FUNCTIONS

Although we cannot obtain realistic estimates for the parameters of the lobster and whelk R-functions, it is possible to formulate equations that give rise to isoclines similar to those of Figure 15.3. For example, the whelk isocline is an inverse linear function of lobster density. Isoclines with this general form are typical for systems in which two species compete for limited resources. An example is the R-function for *inter*specific competition defined by equation (5.6). Assuming that this equation describes the whelk R-function, then

$$R_P = A_P \left(1 - \frac{P + C_{PN}N}{K} \right),$$

(15.1)

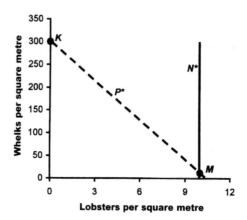

Figure 15.2 Preliminary zero-growth isoclines for lobsters (N^* = solid line) and whelks (P^* = broken line). Note the stable community equilibrium, M, occurs at the position where the two isoclines cross.

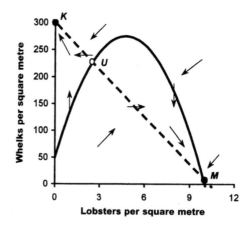

Figure 15.3 Hypothesized zero-growth isoclines for lobsters (N^* = solid line) and whelks (P^* = broken line) showing an unstable community equilibrium at U; the arrows show the direction of population change (vectors). Note that the vectors cross the lobster isocline vertically and the whelk isocline horizontally.

where R_P is the realized per capita rate of change of whelks, A_P is the maximum per capita rate of change of whelks in a given environment, P is the density of the whelk population at the beginning of the interval over which population change occurs, K is the whelk carrying capacity in the absence of lobsters ($K \cong 300$ per square metre), N is the density of lobsters, and C_{PN} is the relative ability of lobsters to compete with whelks for their shared food (note that the time subscript $t - 1$ is omitted from the dynamic variables N and P for simplicity). The zero-growth isocline for the whelk population is obtained by solving equation (15.1) for $R_P = 0$

$$0 = A_P \left(1 - \frac{C_{PN}N}{K} \right).$$

$$P^* = K - C_{PN}N$$

(15.2)

This turns out to be a straight line like that in Figure 15.3, with slope defined by C_{PN} and intercept on the P-axis at K. The slope of the isocline can be estimated by solving equation (15.2) with the estimated values $K = 300$, $P_k^* = 10$, and $N_k^* = 10$

$$10 = 300 - C_{PN} \times 10$$

$$C_{PN} = 29.$$

(15.3)

However, because the whelk isocline does not depend on A_P, the maximum per capita rate of change of whelks, we are unable to estimate this parameter. For the sake of argument assume that $A_P = 2$.

In contrast to the whelk isocline, the parabolic lobster isocline is typical of a prey population (e.g. see Figure 6.5). Using the prey equation (6.5) for the lobster R-function gives

$$R_N = A_N - C_N N - C_{NP} \frac{P}{N + F},$$

(15.4)

with R_N the realized per capita rate of change of lobsters, A_N their maximum per capita rate of change in a given environment, and C_{NP} the predatory impact of whelks on lobsters. Solving equation (15.4) for $R_N = 0$ yields the zero-growth isocline for the lobster population

$$P = \frac{(N + F)(A_N - C_N N^*)}{C_{NP}},$$

(15.5)

which is a parabola intercepting the lobster axis at the lobster carrying capacity (near M), and the whelk axis at the maximum density of whelks among which individual lobsters can survive (N^*, P). Parameters of the lobster isocline were estimated by trial-and-error fitting of equation (15.5) through or near the fixed points M and (N_0^*, P) under the constraint that the two isoclines must intercept to create an unstable point, U. Of course, there are many possible parameter combinations that meet these conditions, one of which is the set $A_N = 2$, $C_N = 0.2$, $C_{NP} = 0.02$, $F = 0.5$ that produced the lobster isocline in Figure 15.3.

15.2.1 Interpretation

The structure of the isoclines deduced for the lobster and whelk populations leads to a biological interpretation of the interaction between these two organisms. On the one hand, the whelk isocline is similar to that produced by *inter*specific competition between two different species for shared resources. This suggests that the dominant effect of lobsters on whelks is through competition for shared food rather than predation. This interpretation seems reasonable for the following reasons: although lobsters will eat whelks, they are not a preferred food. In addition, large thick-shelled whelks and whelks with bryozoans growing on their shells are immune to rock lobster attack. Finally, although lobsters and whelks are both scavengers on the same animal food, lobsters are much more mobile and can also feed on living animals. Hence, lobsters have a significant competitive advantage over whelks, particularly when dying or dead animal matter is scarce.

On the other hand, the form of the lobster isocline suggests that the main effect of whelks on lobsters is through predation rather than competition. This interpretation also seems reasonable for the reasons stated above. It is also supported by experimental evidence demonstrating that large whelk populations can kill healthy lobsters by mass attack.

Finally, it should be remembered that the isoclines deduced in Figure 15.3 are not the only possibilities. There are certainly other models that can give rise to the equilibrium points observed at Malgas and Marcus Islands. For example, it is possible to develop a non-linear model for two competitors that produces similar equilibrium points. However, the biological reasons for this, and other possible isocline structures, do not seem as plausible as those for the model developed above.

15.3 SIMULATED POPULATION DYNAMICS

It is now possible to simulate the dynamics of hypothetical lobster and whelk populations by employing the R-functions of equations (15.1) and (15.4) and the stochastic forecasting equation (3.13) (Figure 15.4). When both populations start at low densities in a constant environment ($s_N = s_P = 0$) (Figure 15.4, top-left), the whelk population tends to increase rapidly but is eventually brought back to the community equilibrium if sufficient lobsters are present. If the lobster population is too small, however, whelks can reach densities that drive the lobsters to extinc-

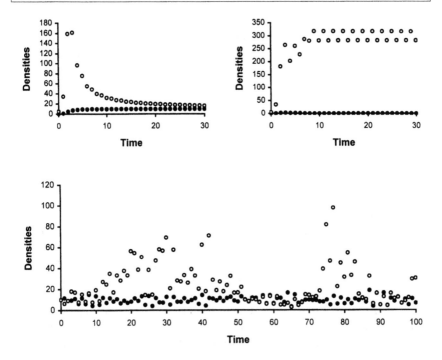

Figure 15.4 Simulation of lobster and whelk population dynamics with the R-functions defined by equations (15.1) and (15.4) and the step-ahead forecasting equation (3.13). Top-left, deterministic simulation starting from the right of the separatrix $(N_0, P_0) = (0.2, 5)$ per square metre, showing an asymptotic approach to the community equilibrium $(N_m^*, P_m^*) = (10, 10)$. Top-right, deterministic simulation starting from the left of the separatrix $(N_0, P_0) = (0.1, 5)$ per square metre, showing an oscillatory approach to the whelk carrying capacity and local extinction of the lobster population $(N_k, P_k^*) = (0, 300)$. Bottom, stochastic simulation with initial conditions at the community equilibrium $(N_0, P_0) = (10, 10)$ and standard deviations $s_N = s_P = 0.2$.

tion and the whelk population eventually fluctuates around its carrying capacity, K (Figure 15.4, right).

Stochastic simulations show that the community equilibrium is very robust to external disturbances (Figure 15.4, bottom). Extinction of one or the other species was only observed if very large standard deviations were used. For example, the extinction of lobsters was only observed when standard deviations exceeded 0.5 $(s_N \geq 0.5)$, which represents a standard deviation 25% of the maximum per capita rate of change. Even larger disturbances on the whelk population $(S_P \cong 1)$ were required to cause the extinction of lobsters. Simulations also show that lobster populations are much less variable under similar disturbance regimes than whelk populations, the latter sometimes straying far from the community equilibrium (Figure 15.4, bottom). The simulation experiments suggest that the lobster-dominated community is only likely to move to the whelk-dominated situation following catastrophic disturbances of the lobster population.

15.4 MANAGEMENT IMPLICATIONS

There are two important questions relevant to the management of the lobster fishery off the west coast of South Africa. The first concerns the level of lobster fishing that can be sustained without driving rock lobsters to local extinction by the combined action of harvesting and whelk predation. A tentative answer to this question can be obtained from the isocline diagram (Figure 15.5). Here a harvest strategy is simulated that maintains a constant density of 2 lobsters per square metre. Under these conditions, whelk densities gradually increase until the system crosses a *separatrix* close to the descending arm of the lobster isocline (the sepa-ratrix will be identified later). Once this happens, the lobster population is driven automatically to extinction by whelk predation, even if harvesting is discontinued. Notice that harvesting lobsters to a density of 3 or more

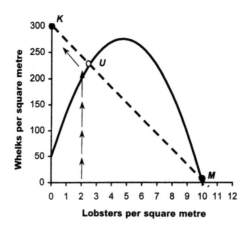

Figure 15.5 Qualitative assessment of a harvesting strategy that maintains lobsters at about 2 per square metre for a number of years. Note that if the lobster population is artificially maintained at a constant density, the whelk population will grow until it reaches its own isocline, or until it crosses a separatrix near the descending arm of the lobster isocline. Once the separatrix has been crossed, the whelk population automatically evolves to its own carrying capacity, K, and lobsters are driven locally extinct by whelk predation, even if harvesting is discontinued.

per square metre results in stable community with a density of whelks given by equation (15.2). For example, if lobsters were artificially maintained at a density of 6 per square metre by fishing, a community equilibrium would be created at $M = (N_m^*, P_m^*) \cong (6, 126)$. It is important to realize, however, that the closer the community equilibrium is to the unstable threshold, U, the greater the probability that random environmental disturbance will push it across the unstable point U and into the domain of the whelk-only equilibrium, K. Because we do not know the exact location of the unstable point, is seems wise to employ a conservative harvest strategy.

The second management question concerns the re-establishment of lobster fisheries in areas dominated by whelks. It is here that the identity of the separatrix is most useful. A separatrix is a line defining the basins of attraction to two (or more) different equilibrium points (see Section 10.2.2.4). The approximate location of this line can be determined by simulation (Figure 15.6). When the initial densities of lobsters and whelks (the coordinate N_0, P_0) are to the left of this separatrix, the system will evolve towards the whelk-only equilibrium, K, while if they are to the right, the system will evolve towards the community equilibrium, M. Hence, the lobster fishery can only be maintained or re-established if the coordinates of the system are kept by management actions to the right of the separatrix.

The re-establishment question can now be approached as follows:

1. *Importing lobsters.* Given that the whelk carrying capacity is about 300 per square metre, then one would need to import about 3 lobsters per square metre to move the system coordinate to the right of the separatrix and re-establish a viable lobster population (Figure 15.6). Assuming that the area controlled by whelks around Marcus Island is about 100,000 square metres, then at least 300,000 lobsters would have to be released simultaneously over the area in order to re-establish the fishery. This may be a difficult or impossible task.
2. *Reducing whelks before importing lobsters.* An alternative strategy is to reduce the whelk population before introducing lobsters into the area. The greater the reduction of the whelk population, the lower the density of lobsters needed to attain the same ends. For example, if the whelk population can be reduced to 10 per square metre, only 0.2 lobsters per square metre would be required, or a total of around 20,000 lobsters over the entire area.

Of course, a discussion of rock lobster management based on these models is largely hypothetical because of uncertainty in the feedback structure of the model as well as the values of its parameters. Even so, the model can provide the manager with a feel for the system and its behaviour under management, something that would be impossible without some sort

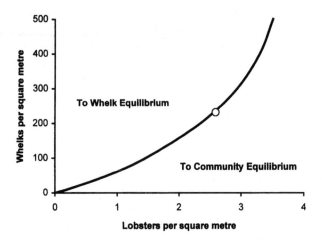

Figure 15.6 Identification of the separatrix defining the basins of attraction to the whelk-only equilibrium, K, and the community equilibrium, M. The position of the separatrix was determined by simulation: given initial densities of whelks and lobsters, the trajectory of population change was calculated with equations (15.1), (15.4) and (3.12). If the trajectory evolved towards K, the density of lobsters was increased slightly and another trajectory calculated, repeating this process until evolution towards M was observed. When the initial trajectory was observed to move towards M, the density of lobsters was decreased slightly and another trajectory calculated and so on until the approximate coordinate (N_S, P_S) on the separatrix, S, was obtained for a particular density of whelks. This convergence procedure was then repeated for another 10 fixed densities of whelks and a smooth line was drawn through the 11 coordinates. The open circle is the unstable community equilibrium, U, which lies on the separatrix.

of theoretical analysis and modelling. Finally, the hypothetical isocline structure for the lobster–whelk system opens up the opportunity for experimental studies or adaptive management strategies. For example, controlled experiments could be designed, perhaps on small isolated islands or enclosures, in an attempt to define the actual basins of attraction of the two equilibrium points and the location of the unstable separatrix. If this was not possible, perhaps the same information could be obtained by managing the fishery intelligently, say by varying harvesting intensity in time and space, or by reducing whelk populations and/or stocking lobsters.

15.5 EXERCISES

1. Use the model for the lobster–whelk interaction to try and construct a qualitative single-species R-function for the lobster population. To do this, set the whelk population constant and then calculate R_N with

equation (15.4) for different lobster densities (say 0–12 lobsters per square metre). Plot the R-function on some graph paper. Repeat this for smaller or larger constant densities of whelks to see the R-function change with whelk density. Note that the lobster single-species R-function should have an extinction threshold.

2. Construct the qualitative single-species R-function for the whelk population. Note that the whelk R-function must have an escape threshold.

16	# Human beings: retrospect and prospect

If only Malthus, instead of Ricardo, had been the parent stem from which nineteenth-century economics proceeded, what a much wiser and richer place the world would be today. (J. M. Keynes, 1951)

The objectives of this, the final chapter, are two-fold: first, to summarize the ideas developed in this book and put them into some kind of historical perspective, and second to show how these ideas can be used to investigate the growth of the human population on the planet Earth. This approach seems natural, for some of the great ideas about population dynamics have come from thinking about human beings, and it seems only natural for modern humans to think about their own future.

16.1 THE "GREAT IDEAS"

Although many have contributed to the contemporary theory of population dynamics, it is possible to recognize a few "great ideas" that, in my opinion, laid the conceptual corner-stones on which the theory rests. These ideas and their creators are briefly discussed in cronological order.

16.1.1 Malthus 1798; Verhulst 1838

Two hundred years ago the British clergyman Thomas Malthus published his monumental and controversial *Essay on the Principle of Population*. Malthus was an economist as well as a preacher, and was, therefore, interested in the relationships between the expanding human population and the economic and spiritual well-being of individuals. Like others before him, including Giovanni Botero and Benjamin Franklin, Malthus clearly understood the enormous power of the first principle of population

dynamics, the potential for self-replicating populations to grow geometrically. This was also the beginning of the "economic connection" (see later)
for, as an amateur economist, Malthus was well aware of geometric
progressions in compound interest savings accounts. The first principle is
sometimes known as the "Malthusian law" and the growth parameter (R)
the "Malthusian parameter".

Malthus also understood that resources were, in the absolute sense,
finite and that the realities of exponential growth and finite resources
would lead to a "struggle for existence" amongst the members of the
expanding population, and that the resulting "vice, famine and warfare"
would eventually limit population growth. In other words, Malthus was
aware of the third principle of population dynamics. Forty years later the
Belgian mathematician P.-J. Verhulst put these ideas together in his
logistic equation[17], the first mathematical model for population growth in
a limiting environment, and the basic equation underlying many of the
mathematical ideas developed in this book.

However, Malthus was wrong in thinking that the human population
would soon overwhelm the resources available to it and suffer the drastic
consequences of the third principle. The reason for his mistake was a
lack of understanding of the second principle of population dynamics (the
idea of "increasing returns" had yet to penetrate economic thinking).
Malthus thought that food would be the major limiting resource and
believed that food could only increase as a linear function of time. This
led him to predict that the time was swiftly approaching when the exponentially growing human population would come up against the limits of
its food supply. If Malthus had understood the second principle, however,
he would have realized that humans co-operate with each other in
acquiring knowledge and technology, and that this co-operative venture
would enable them to increase food production *as a function of human
population size*. In other words, food production would also increase
geometrically with time and that, because of this, the "struggle for existence" would be postponed. The failure of Malthus' theory to predict the
fate of the human population is often used by his critics as proof that
the theory is false. Nothing could be more dangerous and further from
the truth! Malthus' theory was correct, as far as it went, but his theory
was incomplete. Despite this, Malthus' essay was the first clear articula-
tion of the first and third principles and, because of this, it seems
appropriate to recognize him as the "father" of population dynamics.

16.1.2 Liebig 1840; Blackman 1905

The next major contribution to population dynamics theory came through
experimental research in the agricultural sciences. In 1840, the German
agricultural chemist Justus Liebig published a book entitled *Chemistry and*

its Application to Agriculture and Physiology. Liebig was interested in the effects of environmental factors on the growth of plants, and found, in his experiments, that plant growth and yield tended to be limited by the nutrient that was in shortest supply. In 1905 Blackman coined the term "limiting factor" to describe this phenomenon. What this means in Malthusian terms is that the *struggle for existence* will be most intense for the resource (or group of interdependent resources) in shortest supply, and that this struggle will result in increased deaths and/or decreased fertility that will, eventually, limit population growth. From these ideas springs the fifth principle of population dynamics, a principle that simplifies and organizes our thinking about the multitude of potential factors affecting population change. The fifth principle states that, of all the factors acting on the population growth rate, only one of them, or a group of interdependent factors, will actually set the limits that determine the equilibrium density of the population.

Of course, it is difficult to see how the fifth principle relates to human population dynamics because it only applies, in the strict sense, near equilibrium, and the transient human population is definitely not in this state. Malthus believed that food would be the limiting factor, but most demographers nowadays feel that population growth will be halted by the "demographic transition"[46]. As we will see later (Section 16.6) there are many other potential limiting factors, but it is difficult or impossible to determine which of them will limit the human population while it is in a transient state.

16.1.3 Lotka 1925; Volterra 1926; Elton 1948; Hutchinson 1948

The next of the "great ideas" involves the notion of regular population cycles and the fourth principle of population dynamics. The notion of population cycles seems to have begun with the mathematical models of the American chemist Alfred J. Lotka and the Italian mathematician Vito Volterra. Working independently on identical models for the interaction between predator and prey populations, Lotka and Volterra demonstrated that cyclic population fluctuations are an intrinsic property of the model[23]. Then, in 1948, came two important publications – the book *Voles, Mice and Lemmings* by Oxford professor Charles Elton and the paper *Circular Causal Systems in Ecology* by Yale professor Evelyn Hutchinson[22]. Elton's book documented the common occurrence of regular population cycles in small rodents, while Hutchinson provided a general explanation for the causes of cyclic dynamics in ecological systems. Hutchinson argued that circular causal pathways can be created when populations affect the properties of their environments, and that time delays in these feedback loops can cause spectacular cycles of abundance – the "boom and bust" of algal blooms and outbreaks of mice, voles, lemmings, and insect pests (e.g. the larch budmoth in Figure 9.2).

Of course, the relevant question for human beings is whether they will be subjected to the drastic consequences of the fourth principle. What is clear is that the expanding human population is causing significant changes to its environment: rivers are being dammed, forests and fresh water depleted, air and water polluted, arable land turned into cities. The burning of fossil fuels has increased the so-called "greenhouse effect" and the phenomenon of "global warming". Chemicals released into the atmosphere cause "ozone depletion" and increasing ultra-violet radiation. What is not clear is whether these environmental changes will feed back to significantly affect the reproduction and/or survival of future human beings and, thereby, evoke the fourth principle and its dire consequences. We will discuss this possibility further in Section 16.6.

16.1.4 Allee 1931; Boserup 1965

The most recent of the "great ideas" came from the study of animal behaviour and, in particular, from an area of biology that is nowadays called sociobiology. In 1931, Professor W. C. Allee of the University of Chicago published a book entitled *Animal Aggregations: A Study in General Sociology* in which he documented the widespread occurrence and importance of co-operative interactions amongst living organisms. Although the importance of social interactions in animal ecology had been recognized previously, as for instance in A. V. Epinas' book *Des Societies Animales*, it was Allee's work that documented in great detail the significance of incidental (non-adapted) co-operative interactions and aggregations in animal populations. At the same time, economists and anthropologists had also been thinking about the potential effects of human aggregations on agricultural and industrial development. In 1965, the Danish economist Ester Boserup developed these ideas into a comprehensive theory in her book *The Conditions of Agricultural Progress*. In this book, Boserup argued that the rate of technological change, and through it food production, should increase in direct relation to the density of the human population. If this proposition is true, then the human population has probably been under the influence of the second principle of population dynamics for at least three centuries. Thus, unlike Malthus, we are now able to understand why the per capita rate of change of human beings can increase as a function of population density (Figure 10.1) and how this can cause the population to grow at a hyper-exponential rate (Figure 9.2).

16.2 THE MODERN SYNTHESIS

Throughout this book I have attempted to convince the reader that, contrary to many claims in the ecological literature, there exists a set of principles, relationships and causal processes that explain the complete range of dynamic behaviour observed in natural populations. In other words, I have argued for a comprehensive and logical *theory of population dynamics* – what is sometimes called the *modern synthesis*. Although this synthesis rests upon the "great ideas" summarized above, literally hundreds of scientists have been involved in the collection and analysis of data, and the integration of these data and ideas to form the modern synthesis. Schools of population dynamics grew at a number of well-known research institutions during the last half-century. The "great debate" of the 1950s (and still continuing) over the relative importance of exogenous (density-independent) and endogenous (density-dependent) factors brought population dynamics theory to the forefront of ecological thinking[12]. Theoretical and experimental studies of competition and predator–prey dynamics expanded our understanding of the interactions within and between populations and their dynamic consequences. Long-term studies on a number of economically important forest insects provided the data to challenge the emerging theory, and generated new methods for analysing and interpreting those data. Out of all this has evolved a comprehensive and general theory of population dynamics, the modern synthesis presented in this book.

Underlying the modern synthesis, of course, are the five principles of population dynamic. The first principle stands as a universal "law" because it applies to all populations all of the time and in all places, given that the per capita rate of change is known. The second and third principles determine the relationship between population density and the per capita rate of change for specific organisms living in specific places at specific times and, thereby, the form of the R-function and the consequent properties of stability and instability (the stability landscape) of the population system. The fourth principle determines the details of the dynamic behaviour in the vicinity of a stabilizing equilibrium (the order of the dynamics), and the fifth helps to identify the proximal biological factors that set the limits to population growth and the level at which stability is attained. Taken together, the five principles have the potential to explain and predict the dynamics of any population of living organisms. In the case of the human population, for example, the historical dynamics seem to be adequately explained by the first and second principles alone. However, the third and fourth principles can possibly be evoked in the future and this may give us cause for concern (see Section 16.6).

16.3 POPULATION ANALYSIS

The key to understanding the population dynamics of any given species is, of course, the R-function, for if this relationship between the per capita rate of change and population density can be defined mathematically, it becomes possible to describe the stability properties of the population system and to predict its future dynamics. Mathematical modelling of real population data began in the 1920s when demographers first attempted to describe and predict the growth of the human population with the logistic model[17]. Then various theoretical models, including the logistic equation, were used to describe the dynamics of a number of experimental animal populations, among them the southern cowpea weevil[31]. However, it was not until the advent of long-term field studies of economically important insects that ecologists first began to think about the analysis and interpretation of field data. Most of the early studies involved the development of "life tables" and the use of various methods to determine the *key factor* or factors responsible for the observed population fluctuations from such data. Life tables still remain a major way of studying the dynamics of field populations, and their analysis and use is described in many modern textbooks[47].

The problem with life tables is that they require intensive sampling of the population of interest, as well as the factors affecting it, for many years and many times each year. Because of this, the cost of life table studies is prohibitive for most resource and pest management agencies. In recent years a statistical approach called "time series analysis" has been adopted and adapted in an attempt to serve the needs of applied ecologists and population managers[48]. The time series approach requires much less intensive sampling, and can often be applied to data that managers acquire in their normal operations, such as a harvest record or periodic census. This is the approach taken in this book, written for the student of applied population ecology.

16.4 THE ECONOMIC CONNECTION

Throughout this book analogies have arisen between the ideas of population dynamics and economics. This should not be too surprising. Not only was Robert Malthus, the father of population dynamics, an economist rather than an ecologist, but the human economic system is driven by population-level processes, including co-operation and competition (the second and third principles being analogous to the economic "laws" of increasing and diminishing returns). The parallels do not end here. Economic systems are driven by supply and demand relationships in accordance with the general notions of pricing. In this sense, the price paid for a good is directly proportional to the demand/supply ratio; i.e.

$$p = \frac{kN}{S},$$
(16.1)

where k is the individual demand for a good or commodity, S the supply of the good, and N the number of individuals in the population of consumers. This equation states that the higher the total demand for a good (total demand = kN), the greater the price will be, all else being equal. Notice that the price can be lowered by increasing the supply, S, or decreasing the per capita demand, k, or the population size, N.

In the ecological context, p can be considered as the energetic price paid by an organism to obtain the k units of energy required to meet its energetic demands. From this it follows that the energy available for maintenance, growth, and reproduction is

$$k - p = k - \frac{kN}{S} = k\left(1 - \frac{N}{S}\right).$$ (16.2)

Now if w units of energy are required for each unit increase in the realized per capita rate of change, R, then the R-function for the population will be

$$R = wk\left(1 - \frac{N}{S}\right),$$ (16.3)

where $wk = A$ is the maximum per capita rate of change attained when all energy demands are met. This equation is identical to the linear single-species logistic model incorporating the third principle [equation (5.5)] and the two-species logistic model incorporating the fourth principle [equation (6.4)]. Hence, the logistic equation serves to unify the disciplines of economics and ecology.

Of course, there are also fundamental differences between human economics and the economy of nature. Perhaps the most obvious is that the currency of human economics is money while nature's currency is energy. As we shall see later, this divergence of human beings from nature's way is a major contributor to the human dilemma. Be that as it may, it is apparent that human beings are, in the long run, subject to the laws of nature just like any other organism.

16.5 THE EVOLUTIONARY CONNECTION

Perhaps the greatest idea of all in biology was Charles Darwin's *theory of evolution by natural selection*. What is generally less appreciated is that Darwin's theory was based, to a large extent, on Malthus' concept of a "struggle for existence" – the third principle of population dynamics. Simply stated, Darwin proposed that those individuals better adapted to obtain scarce resources, and to escape being used as resources by other organisms, tend to survive and to pass on their adaptive traits to their offspring, while those not so well adapted perish before reproducing. Through this process of natural selection, Darwin argued, species become better adapted to survive and reproduce in a given environment and, under conditions of reproductive isolation, can evolve into new species.

Evolution and population dynamics are intimately intertwined. The evolutionary process gives rise to new traits. New traits change the parameters of the R-functions and the resultant dynamics of the interacting populations. Population dynamics drive the evolution of new traits through the "struggle for existence". And so the mutually causal ⁺feedback loop between evolution and population dynamics is closed. Each process drives the other in an unending progression of adaptation, speciation and changing population dynamics.

In the case of human beings, however, the connection between population dynamics and evolution may have been broken. Unlike other animals, modern humans can control their own reproduction and, thereby, the transmission of adaptive traits and the continuation of the genetic line. In addition, they also control the reproduction and genetics of their domesticated plants and animals. The consequences of this interference with the "natural" evolutionary process is open to speculation, but even speculation is difficult while the human population continues to grow at its current rate. Perhaps human evolution will resume again if and/or when the third and fourth principles rear their forbidding heads and the "struggle for existence" resumes once more.

16.6 THE HUMAN FUTURE

And so we return to the population dynamics of human beings. It should be apparent by now that the long-term behaviour of a population cannot be understood or predicted if it has not been observed near equilibrium. There should also be no doubt that the human population must, at some time, cease growing on this planet of finite dimension (i.e. space, if nothing else, must be a limiting factor). But when this will occur and what process will bring it about cannot be predicted with any certainty while the human population is in a transient state (i.e. it is difficult or impossible to apply the fifth principle under these conditions). The current thinking amongst demographers is that stabilization will result from the "demographic transition" – the voluntary reduction of family size as technology and industrialization spread to all nations[46]. This transition can be viewed as a special case of the third principle operating within the human economic system. In this sense, individuals compete to obtain both essential resources (food and a place to live) and non-essential (luxury) resources, but the currency for obtaining both is money. Hence, money is really the common resource for which human beings compete.

In modern industrial societies, infant mortality is low, and the cost of raising children high. Families generally become more competitive in the struggle to obtain financial resources if both spouses work. For this reason, it seems only natural that family size should shrink as human populations become more dense and industrialized. Notice that the driving force is still the third principle, the competitive struggle to obtain resources, in this case money. The outcome is the "demographic transition", the voluntary limitation of family size as populations become more dense and industrialized.

The trouble with this application of the third principle is that there is no upper limit to the individual demand for wealth. Unlike nature's economy, where individual demands are finite (organisms can only eat so much food), the demand for money seems to be insatiable. No wonder that wealth becomes more and more concentrated in fewer and fewer families as the rich pass on their wealth to fewer and fewer children. Of course, they also pass on fewer and fewer genes, and may thereby loose out in the long "evolutionary" struggle.

What does all this mean to the future of the human population? The actual growth of the human population over the last 350 years is illustrated by the data points (census estimates) in Figure 16.1. The first line in this figure is the growth trajectory predicted by the R-function

$$R = AN_{t-1}, \tag{16.4}$$

with the value of A estimated by fitting the growth equation to the data. The assumption that R increases linearly with population density was based on Figure 10.1, and implies that the second principle acts linearly on the human per capita rate of change. This model predicts that the human population will reach an incredible 51 billion people by 2025 and an unimaginable number shortly afterwards – the "doomsday prediction" of some simple models of human population dynamics[39].

Of course, equation (16.4) cannot be strictly true because, as we saw in Chapter 4, the per capita birth rate must have a maximum value related to the reproductive potential of the species (e.g. see Figure 4.2). In fact the data suggest that the human population trajectory after 1985 is better described by a linear (geometric) model than the hyper-geometric model; i.e. the growth trajectory becomes approximately linear on the logarithmic scale over the period 1985–1995 (Figure 16.1). This indicates that the per capita rate of change became constant in the mid-1980s. Assuming that the human population will continue to increase geometrically yields the second projection in Figure 16.1. In this case the population is predicted to grow to a much more reasonable 10 billion by the year 2050.

The next thing to consider is the possibility that the "demographic transition" (or some other factor) acts in accordance with the logistic equation to cause R to fall as population density rises. In this case an appropriate R-function may be

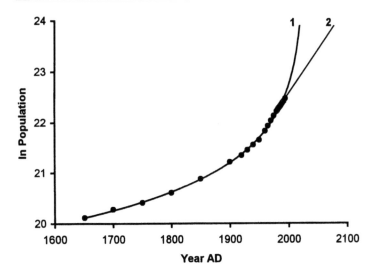

Figure 16.1 Line 1, convergence fit[32] of the model $N_t = N_{t-1}e^R$ with $R = AN_{t-1}$ to the human time series, with parameter estimate $A = 0.00000487$. Line 2, after 1985 the population growth is assumed to grow in accordance with the first principle (i.e. as a geometric rather than a hyper-geometric progression) and so the data from 1985 to 1995 are fitted to the model $N_t = N_{t-1}e^R$ with $R = 0.0172$ (see reference 39 for data source).

$$R = AN_{t-1}\left(1 - \frac{N_{t-1}}{K}\right), \tag{16.5}$$

with K the equilibrium density, or carrying capacity of the planet. Fitting this model to the human time series gives the prediction shown in Figure 16.2 (line 1). Under this scenario, the human population would stabilize at 20 billion by the middle of the 21st century. Alternatively we could adopt the exponential model after 1985, in which case the appropriate R-function would be

$$R = A\left(1 - \frac{N_{t-1}}{K}\right). \tag{16.6}$$

When this model is fitted to the human data it predicts that the population will reach 10 billion by the year 2050 and will stabilize at 12 billion by 2200 (line 2 in Figure 16.2).

In the models above it was assumed that an increase in population size was immediately fed back to affect reproductive rates through the "demographic transition". This assumption causes the population to approach equilibrium smoothly (asymptotically) because the fourth principle is not in operation. It seems equally likely, however, that a time delay could be

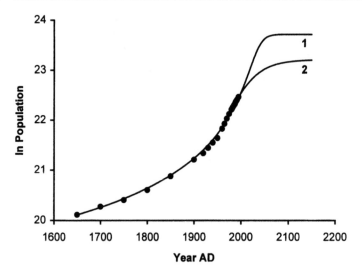

Figure 16.2 Line 1, convergence fit of the model $N_t = N_{t-1}e^R$ with $R = AN_{t-1}$ $(1 - N_{t-1}/K)$ to the human time series with parameter estimates $A = 0.0000052$, $K = 20,000,000,000$. Line 2, convergence fit of the model $N_t = N_{t-1}e^R$ with $R = A(1 - N_{t-1}/K)$ to the human time series with parameter estimates $A = 0.03$, $K = 12,000,000,000$.

involved in feedback acting through the demographic transition; i.e. it may take many years for families to reduce their size following increases in population density and industrialization. Models (16.5) and (16.6) can be modified to account for time delays (d) in the demographic feedback by replacing $t - 1$ with $t - d$. The fit of this model (16.5) with time delays of 10 and 20 years gives the predictions shown in Figure 16.3. When the time delay is 10 years, the population stabilizes at about 10 billion people after minor oscillations (line 1 in Figure 16.3). However, a time delay of 20 years produces a massive population crash to about 70 million people in the year 2080 (line 2 in Figure 16.3). Unfortunately it is not possible to use model (16.6) because there are only 10 years of data over the period where the geometric growth assumption holds (1985–1995). However, in the absence of the destabilizing second principle, we would expect a more moderate growth trajectory.

Although the scenarios simulated above must be considered highly speculative, they do cover the gamut of possible population dynamics. First, the human population could continue to grow at a hyper-geometric or geometric rate (Figure 16.1), although how this could be accomplished is difficult to imagine. Even colonization of new planets at the speed of light could not sustain this kind of growth. Second, the population may slow down gradually because of a rapid demographic transition (or some

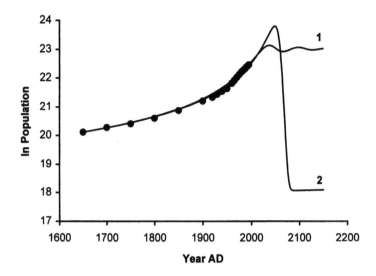

Figure 16.3 Line 1, convergence fit of the model $N_t = N_{t-1}e^R$ with $R = AN_{t-1}$ $(1 - N_{t-d}/K)$ to the human time series with parameters $A = 0.00000555$, $d = 10$, $K = 10,000,000,000$. Line 2, same as above but with $d = 20$ and $K = 8,000,000,000$.

other feedback process that operates quickly) and finally attain equilibrium asymptotically (Figure 16.2) or with damped oscillations (Figure 16.3, line 1). Under these scenarios, it seems likely that the human population will eventually stabilize at between 10 and 20 billion people. On the other hand, it is also possible for the population to follow the more ominous "boom and bust" trajectory illustrated by line 2 in Figure 16.3. This kind of dynamics results from long time delays in the feedback acting to oppose population growth, time delays that result from mutually causal interactions between the population and its environment (the fourth principle). In the case of demographic transition theory, the change in the environment is associated with urbanization and industrialization. However, the rapidly expanding human population is also inflicting other impacts on its environment as rivers are dammed, forests and fresh water depleted, air and water polluted, and so on. All these effects can potentially feed back to affect the reproduction and/or survival of future human beings and, thereby, evoke the fourth principle of population dynamics. The warning is clear, a "boom and bust" trajectory must be considered a possibility in the human future.

This study of human population dynamics has focused attention on the total or global human population. Others may criticize this approach, preferring to identify many different human populations according to national or continental boundaries. However, our definition of a "population" is based on a spatial resolution large enough to encompass the mobility of the species. Because human beings move freely about the planet carrying genes, diseases, plants and animals (including pests) to all corners of the Earth, we probably have to consider them as belonging to one global population if we wish to apply the principles developed in this book. True, there may be small populations of humans (e.g. bushmen) that are separated from, and uninfluenced by, the outside world, but the effect of these small isolated populations on the overall scheme of things is probably insignificant. This is not to say that the analysis of "local" (in our sense) human population dynamics is not interesting or significant, but rather that the principles and concepts developed in this book may not be strictly applicable to what we call "sub-populations" with unbalanced immigration /emigration rates.

16.7 FINAL WORDS

And so we come to the end of this book, begun more than a decade ago. The primary purpose for writing this book was to support my population dynamics courses, now developed for the World Wide Web under the titles "Population Theory" and "Population Analysis". Because this book was written for students as well as professionals, I have tried to present the basic principles of population dynamics in a clear and logical fashion, and to foster *understanding* rather than memorization.

The second purpose for writing this book was to advance a diagnostic approach to ecological analysis and assessment. I believe that the diagnostic approach is the only practical one for those involved in the management and conservation of real field populations. Rarely will these people have time or money to perform detailed life table studies or complicated field experiments, so the diagnostic approach seems ideal for their purposes. But it would be a mistake to think of this approach as a truely scientific method. It is really a way to obtain objective *opinions* from minimal data rather than for testing scientific hypotheses. It is a method for *generating* hypotheses about the possible causes of observed population fluctuations rather than for testing those hypotheses. As such it can never substitute real science, the objective testing of hypotheses with carefully designed experiments.

References

1. Criticism of retrospective ecology

For a critical appraisal of ecology, not all of which I agree with, see the book

Peters, R. H. (1991) *A Critique for Ecology*. Cambridge University Press, Cambridge.

2. Approaches to modelling ecological systems

The case for mixed theoretical–empirical models has been made by Berryman (1991a, b, 1997), and arguments for and against this approach, or defending other approaches, can be found in Onstad (1991), Logan (1994) and Hess (1996).

Berryman, A. A. (1991a) Population theory: an essential ingredient in pest prediction, management, and policy making. *American Entomologist* 37, 138–142.
Berryman, A. A. (1991b) Good theories and premature models. *American Entomologist* 37, 202–204.
Berryman, A. A. (1997) On the principles of population dynamics and theoretical models. *American Entomologist* 43, 147–151.
Hess, G. R. (1996) To analyze, or to simulate, is that the question? *American Entomologist* 42, 14–16.
Logan, J. A. (1994) In defense of big ugly models. *American Entomologist* 40, 202–207.
Onstad, D. W. (1991) Good models and immature theories. *American Entomologist* 37, 202–204.

3. References to data in Figure 1.1

Data were obtained from:

Human population: United Nations' Demographic Yearbooks.
Blue whales: FAO Yearbook of Fishing Statistics.
Sycamore aphids: Dixon, A. F. G. (1990) Population dynamics and abundance of deciduous tree-dwelling aphids, in: A. D. Watt, S. R. Leather, M. D. Hunter

and N. A. C. Kidd (Eds) *Population Dynamics of Forest Insects*. Intercept, Andover.

Larch budmoth: Baltensweiler, W. and Fischlin, A. (1988) The larch budmoth in the Alps, in: A. A. Berryman (Ed.) *Dynamics of Forest Insect Populations: Patterns, Causes, Implications*. Plenum, New York.

Gypsy moth: Burgess, A. F. and Baker, W. L. (1938) The gypsy and brown-tail moths and their control. *USDA Circular No. 464*.

Lake whitefish: Christie, W. J. (1974) Changes in the fish species composition of the Great Lakes. *Journal of the Fisheries Research Board of Canada* 31, 827–854.

4. Individual-based models

With the advent of supercomputers and parallel processors, scientists have been modelling populations by considering each individual separately (see, e.g. DeAngelis and Gross (1992)). This endeavour has provided us with important theoretical information about the kinds of dynamic behaviour to be expected from interactions between the many individuals in a population. However, these theoretical exercises do not solve the practical problem of how to measure the characteristics of individuals in natural field populations, where it is impossible to know the location, genetic properties, ancestry, or inclination of each organism.

DeAngelis, D. L. and Gross, L. J. (Eds) (1992) *Individual-based Models and Approaches in Ecology*. Chapman and Hall, New York.

5. Spatial scale

The problem of spatial scale is discussed at length by Allen and Hoekstra (1992) and briefly by myself (Berryman, 1987), where I note that some ecologists seem to choose spatial dimensions rather arbitrarily, or with an ulterior motive.

Allen, T. F. H. and Hoekstra, T. W. (1992) *Towards a Unified Ecology*. Columbia University Press, New York.

Berryman, A. A. (1987) Equilibrium or nonequilibrium: is that the question? *Bulletin of the Ecological Society of America* 68, 500–502.

6. Sampling methods

Details of methods for sampling populations of biological organisms are beyond the scope of this book. Excellent treatments can be found in:

Morris, R. F. (1955) The development of sampling techniques for forest defoliators, with particular reference to the spruce budworm. *Canadian Journal of Zoology* 33, 225–294.

Southwood, T. R. E. (1978) *Ecological Methods*, 2nd Edn. Chapman and Hall, London.

7. Per capita death rates

It is actually a bit more complicated than this because per capita death rates, being counted over a finite interval of time, must include the death of those organisms alive at the beginning of the time period (the parents) and those born during the time period (the offspring). Hence, the total death rate per capita is actually $D_A + D_Y B$, where $0 \le D_A \le 1$ is the per capita death rate of adult parents that were alive at the beginning of the time period, and $0 \le D_Y \le 1$ is the per capita death rate of young born during the time period. In this respect the *total* change in population density over a time interval is really

$$\Delta N = N_{t-1} - D_A N_{t-1} + B N_{t-1} - D_Y B N_{t-1}$$

or, in plain words,

total change = original parents – dead parents + total births – dead offspring.

In order to maintain simplicity, however, we consider B in the text to be the offspring that are alive at the end of the time period, what is sometimes called *recruitment*, i.e. B in the text is really $B - D_Y B = B(1 - D_Y)$. Hence, the death rate D of the text is the same as D_A above, and so only applies to the original parents.

8. Sensitivity to initial conditions

The phenomenon of sensitive dependence can be illustrated as follows: suppose we have a population of N_0 individuals with a constant rate of change whose growth we wish to predict into the future. However, suppose that our estimate of the density of the population has a small error x_0. The true population will grow according to the equation $N_t = N_0 e^{Rt}$ while the predicted population will grow according to the equation $E[N_t] = (N_0 + x_0)e^{Rt}$, and the difference between the real and predicted population will be

$$x_t = N_t - E[N_t] = N_0 e^{Rt} - (N_0 + x_0)e^{Rt}$$

$$x_t = N_0 e^{Rt} - N_0 e^{Rt} + x_0 e^{Rt} = x_0 e^{Rt}$$

In plain words, the error or difference between the actual and estimated initial population will be amplified exponentially with time provided that $R > 0$. For this and other reasons, sensitivity to initial conditions is an extremely important property of population systems. Note that the error will decay exponentially with time when $R < 0$.

9. "Allee effect"

The ecologist W. C. Allee (1931) was one of the first to write extensively on the ecological significance of animal aggregations. The quote that begins this chapter was taken from his book.

Allee, W. C. (1931) *Animal Aggregations. A Study in General Sociology*. University of Chicago Press, Chicago.

10. Behavioural (functional) responses

Solomon (1949) seems to have been the first to separate the response of consumers to the density of their resources into *functional* and *numerical* responses, with the former describing how the consumption rate of individual consumers changes with respect to resource density and the latter how the per capita reproductive rate changes with resource density. I, and some others, prefer the term *behavioural response* because it describes the hunting *behaviour* of the consumer. Holling (1959a) identified three basic types of behavioural responses:

Type 1 (linear) response in which the attack rate of the individual consumer increases linearly with prey density but then suddenly reaches a constant value when the consumer is satiated (top figure).

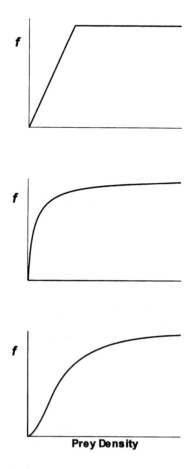

Figure R1 The three basic behavioural or functional responses of predators to the density of their prey with *f* the number of prey devoured per predator per unit of time: top, linear; middle, cyrtoid; bottom, sigmoid.

Type II (cyrtoid) functional response in which the attack rate increases at a decreasing rate with prey density until it becomes constant at satiation (middle figure). Cyrtoid behavioural responses are typical of predators that specialize on one or a few prey species.

Type III (sigmoid) functional response in which the attack rate accelerates at first and then decelerates towards satiation (bottom figure). Sigmoid functional responses are typical of generalist natural enemies which readily switch from one food species to another and/or which concentrate their feeding in areas where certain resources are most abundant[14].

Holling (1959b) derived a mathematical model for the cyrtoid response from experiments in which blindfolded people acted as "predators" searching a table top with their finger tips for a sandpaper disc ("prey"), the so-called disc equation. We can derive the "disc equation" in the following way: define the number of prey attacked by a predator which instantaneously assimilates its prey by

$$f = vN, \tag{R10.1}$$

where f is the number of prey attacked per predator per unit of time (see Figure R1), N is prey density (with $t - 1$ omitted), which is assumed constant over the hunting period, and v is the rate of attack. Now consider the time taken by predators that have to handle, kill and devour its prey before it can search for another. This is called the "handling time" of the predator. If the predator requires t_h time units to handle its prey then, of the total time T exposed to prey, it will spend $T - ft_h$ time actually searching for prey. Thus, the proportion of the total time spent in searching for prey is $(T - ft_h)/T$. If we now insert this modifier into equation (R10.1), we obtain

$$f = \frac{v(T - ft_h)}{T} N = \frac{WN}{N + F}, \tag{R10.2}$$

where $W = T/t_h$ and $F = T/vt_R$. Real (1977) showed that the two major predator responses can be described by the generalized Michaelis–Menton equations of enzyme kinetics

$$f = \frac{WN^q}{N^q + F^q}, \tag{R10.3}$$

where q is the encounter rate between predators and prey needed before the predator reaches maximum efficiency. When $q = 1$ we have a cyrtoid behavioural response (middle figure) and when $q > 1$ the response is sigmoid (bottom figure). Holling (1959a) demonstrated, experimentally, that small mammals feeding on sawfly cocoons have sigmoid behavioural responses due to their switching[14] from one prey to another as prey density increases. Some authors prefer to use "ratio-dependent"[23] behavioural responses in which the quantity of food per predator, or the prey/predator ratio, is substituted for prey density in the equation; for example, N/P is used instead of N in equation (R10.3).

If the behavioural response gives the number of prey killed per predator, then the per capita death rate of the prey attacked by P predators is given by

$$D = \frac{fP}{N} = \frac{WN^{q-1}P}{N^q + F^q}. \tag{R10.4}$$

From this one can see that when $q = 1$ (cyrtoid response), the prey's death rate

$$D = \frac{WP}{F + N},$$ (R10.5)

decreases continuously with prey density (see Figure 4.2, middle). If the birth rate is assumed constant, then the system will be in equilibrium when $B = D$ or

$$0 = B - \frac{WP}{F + E},$$ (R10.6)

where E is the value of N at equilibrium. Rearranging we have

$$E = \frac{WP}{B} - F.$$ (R10.7)

Notice that the unstable threshold gets higher as predator density increases.

When $q > 1$ (sigmoid response), however, the death rate rises with prey density as long as $N < F$, and then declines when $N > F$ (the advanced student should show this numerically). Depending on the parameter values and predator density, the resulting R-function may or may not have equilibrium points but, if it does, there will be two of them, a stabilizing low-density equilibrium (J) and an unstable escape threshold (E), with $J < E$ (see Figure 5.2).

Although Michaelis–Menten–Holling responses are most commonly encountered in the literature, another form was suggested by Ivlev (1955). Rather than orientating his model around prey survival, Ivlev was thinking about how consumers met their energy demands. In his own words,

> ... the actual ration of food eaten by the predator over a certain period of time will, under favorable feeding conditions, tend to approach a certain definite size, above which it cannot under any circumstances increase and which also corresponds to the physiological condition of full satiation.

In other words, the attack rate of a consumer depends on how hungry it is. From this he proposed that the amount of food obtained by an individual consumer (i.e. the behavioural response), changes in the following way with respect to prey density

$$\frac{df}{dN} = \frac{u(W - f)}{W},$$ (R10.8)

where u is the availability or *apparency* of the resource, W is the demand for food, and $0 \leq (W - f) \leq 1$ measures the difference between the amount of food needed and that obtained, or the hunger of the consumer. Watt (1959) takes the same approach but, in addition, argues that the attack rate will also be related to the density of consumers. Simplifying Watt's argument we let the attack rate be inversely related to P so that

$$\frac{df}{dN} = \frac{u(W - f)}{WP},$$ (R10.9)

and, after integrating, we obtain a functional response model derived from the consumer point of view

$$f = v \left(1 - e^{-\frac{uN}{WP}} \right). \tag{R10.10}$$

Notice that this response is "ratio-dependent[23]" because f is a function of N/P rather than N alone.

Holling, C. S. (1959a) The components of predation as revealed by a study of small-mammal predation of the European pine sawfly. *Canadian Entomologist* 91, 293–320.

Holling, C. S. (1959b) Some characteristics of simple types of predation and parasitism. *Canadian Entomologist* 91, 385–398.

Ivlev, V. S. (1955) *Experimental Ecology of the Feeding of Fishes.* Yale University Press, New Haven. (English translation published in 1961.)

Real, L. (1977) The kinetics of functional response. *American Naturalist* 111, 289–300.

Solomon, M. E. (1949) The natural control of animal populations. *Journal of Animal Ecology* 18, 1–35.

Watt, K. E. F. (1959) A mathematical model for the effect of densities of attacked and attacking species on the number attacked. *Canadian Entomologist* 91, 129–144.

11. Prey survival from attack

Thompson (1924) seems to have been the first to propose a model for the survival of a prey population from attack by natural enemies. Thompson was thinking about insect parasitoids attacking insect hosts, but the model can be generalized to other systems with a little caution and imagination. Insect parasitoids lay their eggs on or in an insect host and the young larval parasitoids eventually devour and kill the host. Assuming that parasitoids lay their eggs at random among a population of hosts then, if the populations are large enough, the probability of 0, 1, 2, ... attacks on a particular host is given by the Poisson distribution

$$\Pr(i) = \frac{\bar{X}^i}{i! e^{\bar{X}}}, \tag{R11.1}$$

where $\Pr(i)$ is the probability of exactly i attacks on a host and $\bar{X}$ is the mean number of attacks per host. Now if each parasitoid lays W eggs on N hosts, then the total number of eggs laid by P parasitoids is WP and the mean number laid per host is WP/N. Substituting this quantity for $\bar{X}$ in equation (R11.1), then the probability of a host having no attacks and, therefore, surviving parasitism, is

$$\Pr(0) = \frac{(WP/N)^0}{0! e^{WP/N}} = \frac{1}{e^{WP/N}} = e^{-WP/N}. \tag{R11.2}$$

Now consider the finite population growth equation (3.1), $N_t = N_{t-1}G$, which describes unconstrained population growth (the first principle). Multiplying this equation by the probability of surviving parasitoid attack produces a dynamic model for population growth in the presence of parasitoids

$$N_t = N_{t-1}Ge^{-WP/N}. \tag{R11.3}$$

Converting to natural logarithms yields

$$\ln N_t = \ln N_{t-1} + \ln G - WP/N$$
$$R_N = \ln N_t - \ln N_{t-1} = A_N - WP/N. \tag{R11.4}$$

where $A_N = \ln G$. Allowing for alternative prey in the denominator of the death rate fraction produces an equation identical to text equation (4.9). It is important to realize that the per capita rate of prey survival in equation (R11.3) is converted to a per capita death rate on the logarithmic scale (R11.4).

Thompson's prey survival function underlies many of the models used to study predator–prey relationships. For instance, it becomes the well-known Nicholson–Bailey (1935) model when $W = aN$ (Royama, 1971), and the Michaelis–Menten–Holling model when alternative prey are included (Berryman, 1992). Royama (1971) presents a highly technical treatment of the various models for predator–prey interactions.

Berryman, A. A. (1992) The origins and evolution of predator–prey theory. *Ecology* 73, 1530–1535.

Nicholson, A. J. and Bailey, V. A. (1935) The balance of animal populations. Part 1. *Proceedings of the Zoological Society of London* Part 3, 551–598.

Royama, T. (1971) A comparative study of models for predation and parasitism. *Researches in Population Ecology*, Supplement No. 1, 91 pp.

Thompson, W. R. (1924) La theory mathematique de l'action des parasites entomophages et le facteur du hassard. *Annales Faculte des Sciences de Marseille* 2, 69–89.

12. Density dependence

The negative relationship between population density and the reproduction and survival of individuals is often called *density dependence* in conventional textbooks (in our terminology, density dependence is synonymous with density-induced feedback). However, this term has probably generated more controversy and confusion than any other single issue in ecology. The argument started between the two Australian ecologists A. J. Nicholson and H. G. Andrewartha. Nicholson studied populations of blowflies living in corpses and concluded that density-dependent struggle for food, or intraspecific competition, was the main factor regulating population size. Andrewartha studied populations of rose thrips and found that numbers were largely determined by weather which, of course, is not dependent on the density of thrips. The controversy is largely semantic – a confusion about the meaning of the terms regulation, stability, feedback and so on – but even so it continues to this day. In this book we try to avoid the controversy by not employing the term density dependence. Those wishing to explore the original controversy in more detail should start with the *Cold Spring Harbor Symposium on Quantitative Biology*, Vol. 22 (1957). The debate continues in (among others):

Berryman, A. A. (1991) Stabilization or regulation: what it all means! *Oecologia* 86, 140–143.
den Boer, P. J. (1986) Density dependence and the stabilization of animal numbers. 1. The winter moth. *Oecologia* 69, 507–512.
den Boer, P. J. (1988) Density dependence and the stabilization of animal numbers. 3. The winter moth reconsidered. *Oecologia* 75, 161–168.
den Boer, P. J. (1991) Seeing the trees for the wood: random walks or bounded fluctuations of population size? *Oecologia* 86, 484–491.
Hassell, M. P., Latto, J. and May, R. M. (1989) Seeing the wood for the trees: detecting density dependence from existing life-table studies. *Journal of Animal Ecology* 58, 883–892.
Latto, J. and Hassell, M. P. (1987) Do pupal predators regulate the winter moth? *Oecologia* 74, 153–155.
Wolda, H. (1989) The equilibrium concept and density dependence tests. What does it all mean? *Oecologia* 81, 430–432.

13. Competition for enemy-free space

The notion of competition for "enemy-free space" was first introduced into the literature by Jeffries and Lawton (1984) to describe the "apparent competition" (Holt, 1977) that occurs between different prey species (*inter*specific competition) when they are attacked by a common predator. In this book I also use the concept to include *intra*specific competition among prey to avoid attack by predators. Competition of this kind may arise because of a limited number of hiding places or escape routes, or because aggregations of prey draw the attention of predators and this gives less "enemy-free space" and, hence, more competition.

Holt, R. D. (1977) Predation, apparent competition and the structure of prey communities. *Theoretical Population Biology* 12, 197–229.
Jeffries, M. J. and Lawton, J. H. (1984) Enemy-free-space and the structure of ecological communities. *Biological Journal of the Linnean Society* 23, 269–286.

14. Predator switching and aggregation

The effects of switching and aggregation of generalist predators are discussed in detail by Hassell (1978) and Murdoch and Oaten (1975) and several examples are given of arthropod predators which have sigmoid functional responses. The effect of aggregations of parasitoids on local increases in the density of their insect hosts was beautifully demonstrated in an experiment by Gould and associates (1990). The experiment involved transplanting large numbers of gypsy moth egg masses from outbreak regions into areas with very sparse gypsy moth populations. Within a season, all these local gypsy moth infestations were wiped out by aggregations of insect parasitoids.

Gould, J. R., Elkinton, J. S. and Wallner, W. E. (1990) Density-dependent suppression of experimentally created gypsy moth, *Lymantria dispar* (Lepidoptera:Lymantriidae) populations by natural enemies. *Journal of Animal Ecology* 59, 213–233.

Hassell, M. P. (1978) *The Dynamics of Arthropod Predator–Prey Systems.* Princeton University Press, Princeton, New Jersey.

Murdoch, W. W. and Oaten, A. (1975) Predation and population stability. *Advances in Ecological Research* 9, 1–131.

15. The predator and host resistance "pits"

Morris (1963) and Holling (1965) were two of the first to recognize that predators with sigmoid behavioural responses can create a "predator pit" in the prey's *R*-function, and that this can result in prey being stabilized at very sparse densities by their predators. They also realized that prey can escape predator regulation if changes in the environment cause the stable equilibrium point to vanish, or if the prey population rises above the escape threshold (because of immigration or increased reproduction). This concept was used by Holling and his colleagues (Holling, 1978) to model the dynamics of spruce budworm populations in eastern Canada, and was further expanded by Takahashi (1964), Southwood and Commins (1976) and Berryman (1978). Research on aggressive bark beetles also led to the realization that plant resistance can also create a "resistance pit" similar to the "predator pit" (Berryman, 1978).

Berryman, A. A. (1978) Towards a theory of insect epidemiology. *Researches in Population Ecology* 19, 181–196.

Holling, C. S. (1965) The functional response of predators to prey density and its role in mimicry and population regulation. *Memoirs of the Entomological Society of Canada* 45, 3–60.

Holling, C. S. (Ed.) (1978) *Adaptive Environmental Assessment and Management.* John Wiley, New York.

Morris, R. F. (1963) The dynamics of epidemic spruce budworm populations. *Memoirs of the Entomological Society of Canada* 31, 116–129.

Southwood, T. R. E. and Commins, H. N. (1976) A synoptic population model. *Journal of Animal Ecology* 65, 949–965.

Takahashi, F. (1964) Reproduction curve with two equilibrium points: a consideration of the fluctuation of insect population. *Researches in Population Ecology* 6, 28–36.

16. Derivation of the logistic equation

The third principle of population dynamics can be derived formally by considering a system of randomly distributed organisms drawing resources from a circular *area of influence* of radius *r* around themselves, and competing with neighbouring organisms when their areas of influence overlap; i.e. when the distance to the neighbour is less than 2*r* (Royama, 1992).

This is essentially a simple spatially defined model of sessile organisms, like plants, utilizing a constant supply of a resource, like nitrogen.

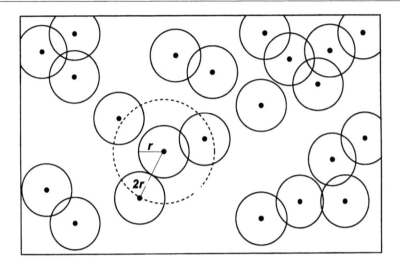

Figure R2 Royama's geometric model of competition between consumers for a fixed density of resources. Each consumer (dots) draws resources from and are a r around itself and competes for resources with other consumers less than $2r$ from it.

Royama (1992, p. 146) analyses the geometric model as follows: if G_i is the finite per capita rate of change when competing with i neighbours, and $\Pr(i)$ is the expected proportion of the population having i overlapping competitors, then the mean finite rate of increase of the population is given by the weighted sum

$$G = G_0\Pr(0) + G_1\Pr(1) + G_2\Pr(2) + \ldots G_i\Pr(i). \tag{R16.1}$$

If the individuals in the consumer population are distributed randomly, then they can be described by a Poisson distribution with mean P, the density of the population (really P_{t-1} but we omit the time subscript for convenience). Under these conditions the number of competitors is also Poisson distributed with mean $4\pi^2 P$. From this the expected proportions are given by the Poisson formula[11]

$$\Pr(i) = \frac{(4\pi^2 P)^i e^{-4\pi^2 P}}{i!}, \tag{R16.2}$$

which can be substituted into the previous equation to yield

$$G = G_0 e^{-4\pi^2 P}\left(1 + \frac{G_1}{G_0} 4\pi^2 P + \frac{G_2}{G_0}\{4\pi^2 P\}^2 / 2! + \ldots \right. \tag{R16.3}$$

Now suppose that the addition of a competitor reduces the average rate of change of an individual by the proportion c ($0 < c < 1$), so that the finite rate of change in competition with i neighbours is $G_i = G_0 c$, and substituting c for G_i/G_0 in the previous equation, we get

$$G = G_0 e^{-4\pi^2 P} (1 + c4\pi^2 P + \{c4\pi^2 P\}^2 / 2! + \dots \tag{R16.4}$$

This equation can then be reduced by Taylor's theorem to

$$G = G_0 e^{-4\pi^2 P} e^{c4\pi^2 P} = G_0 e^{-4\pi^2(1-c)P} = G_0 e^{-C_P P}; \ C_P = 4\pi^2 (1 - c), \tag{R16.5}$$

where C_P is the coefficient of *intra*specific competition between consumers. This equation is, in fact, the well-known Moran (1950) or Ricker (1954) equation. Continuing after Berryman *et al.* (1995), let $R_P = \ln G$ and $A_P = \ln G_0$, so that

$$R_P = \ln G_0 - C_P P = A - C_P P, \tag{R16.6}$$

which is identical to equation (5.2). It is important to realize that the logistic R-function derived in the text and above is the same as the Moran–Ricker model. Now, remembering that the parameter C_P can be expanded to $C_P = V/H$ (Chapter 5), where V measures the demand of the consumer for resources and H is the density of resources, we arrive at the original R-function [equation (5.1)] describing the third principle of population dynamics

$$R_P = A_P - VP/H . \tag{R16.7}$$

In other words, Royama (1992) was able to derive the third principle directly from a mechanistic spatial description of organisms competing for a fixed supply of a limiting resource.

Berryman, A. A., Gutierrez, A. P. and Arditi, R. (1995) Credible, parsimonious and useful predator–prey models – a reply to Abrams, Gleeson, and Sarnelle. *Ecology* 76, 1980–1985.
Moran, P. A. P. (1950) Some remarks on animal population dynamics. *Biometrics* 6, 250–258.
Ricker, W. E. (1954) Stock and recruitment. *Journal of the Fisheries Research Board of Canada* 11, 559–623.
Royama, T. (1992) *Analytical Population Dynamics*. Chapman and Hall, London.

17. Origins of the logistic equation

Stimulated by Malthus' "Essay on the Principle of Population", Verhulst (1838) published what he called a "logistique" equation to describe the sigmoidal growth of population density to carrying capacity. The equation was later rediscovered by Pearl and Reed (1920). Then, Lotka (1925) derived the same equation mathematically, calling it "the law of population growth", and Gause (1934) demonstrated its validity in laboratory experiments. For an excellent review of these historical events see Kingsland (1985). The discrete form of the logistic equation used in this book was first proposed by Cook (1965) but is, in fact, identical to the Moran–Ricker equation[16].

Cook, L. M. (1965) Oscillation in the simple logistic growth model. *Nature* 207, 316.
Gause, G. (1934) *The Struggle for Existence*. Williams and Wilkins, Baltimore.

Kingsland, S. E. (1985) *Modeling Nature*. University of Chicago Press, Chicago.

Lotka, A. J. (1925) *Elements of Physical Biology*. Republished in 1956 by Dover, New York.

Pearl, R. and Reed, L. J. (1920) On the rate of growth of the population of the United States since 1790 and its mathematical representation. *Proceedings of the National Academy of Science* 6, 275–288.

Verhulst, P. F. (1838) Recherches mathematiques sur la loi d'accrossement de la population. *Memoirs de l'Academie Royal Bruxelles* 18, 1–38.

18. Carrying capacity

This is another term that has generated considerable debate and confusion in ecology. However, it serves a useful purpose in defining the maximum density of organisms that a particular environment can *sustain in perpetuity*. It therefore describes the equilibrium population density as determined by the resources available in the region containing the population in question. From this definition, carrying capacities can be exceeded, but only temporarily. For more on the concept of carrying capacity, see:

Catton, W. R. (1980) *Overshoot: the Ecological Basis of Revolutionary Change*. University of Illinois Press, Urbana.

Catton, W. R. (1993) Carrying capacity and the death of a culture: a tale of two autopsies. *Social Inquiry* 63, 202–223.

Focus – *Carrying Capacity Selections*. Carrying Capacity Network, 1325 G Street, Washington, DC 20005–3104, USA.

Harding, G. (1986) Cultural carrying capacity: a biological approach to human problems. *BioScience* 36, 599–606.

Harding, G. (1993) *Living within Limits*. Oxford University Press, New York.

Pulliam, H. R. and Haddad, N. M. (1994) Human population growth and the carrying capacity concept. *Bulletin of the Ecological Society of America* 75, 141–157.

19. Coefficient of curvature

Sometimes called the θ-coefficient, the coefficient of curvature of the R-function is an empirical operator necessitated by real-life observation. It seems to have been first introduced as a modification of Von Bertalanffy's equation by Richards (1959), with methods for fitting it to data being developed by Nelder (1961). The θ-R-function has been applied extensively by Michael Gilpin and his associates (see Thomas *et al.* (1980)).

Nelder, J. A. (1961) The fitting of a generalization of the logistic curve. *Biometrics* 17, 89–110.

Richards, F. J. (1959) A flexible growth function for empirical use. *Journal of Experimental Botany* 10, 290–300.

Thomas, W. R., Pomerantz, M. J. and Gilpin, M. E. (1980) Chaos, asymmetrical growth and group selection for dynamical stability. *Ecology* 61, 1312–1320.

20. Cellular automata

The study of cellular automata has fascinated both school children and professors of mathematics since the introduction of the "game of life" by Conway (1970). This game, which has similar rules to our population game, exhibits a confounding array of spatio-dynamical behaviour. The "game of life" program is available in the *PAS* (see Preface) games module *PG1*. Cellular automata games are also explored in detail in the book by Eigen and Winkler (1981) and can be played on the World Wide Web at the site <http://home.earthlink.net/~hilery/life> or by searching the web for "game of life". Population games are also easy to program and can also be viewed in the *PAS* lesson modules *PL1*, *PL2*, and *PL3*, and the games modules *PG1* and *PG3*.

Conway, G. H. (1970) Mathematical games. *Scientific American* 223(4).
Eigen, M. and Winkler, R. (1981) *Laws of the Game: How the Principles of Nature Govern Chance.* Alfred A. Knopt, New York.

21. Chaos

Chaotic dynamics are characterized by their *sensitive dependence on initial conditions*[8]; i.e. the pattern of fluctuation or spatial arrangement depends on the starting number and spatial arrangement.

> As a whole, the most significant advantage we can gain from mutual causal (feedback) models is that simple rules may generate complex patterns, and that the patterns can greatly vary as the initial conditions vary. (M. Maruyama, 1968)

The concepts of sensitive dependence and complex pattern from simple feedback systems are beautifully illustrated by the "Game of Life"[20] which shows how patterns of growing cellular automata populations depend critically on their starting arrangement.

In chaotic systems, small differences in initial conditions, x_t, diverge exponentially

$$x_t = x_0 e^{\lambda t}, \tag{R21.1}$$

provided that the *Lyapunov exponent* $\lambda > 0$. As we saw in Chapter 3, deviations in the starting states of exponentially growing populations[8] also diverge according to this same equation with $R = \lambda$. By this definition, ⁺feedback due to exponential growth is a chaotic process. Not so obvious, perhaps, is the fact that ⁻feedback processes can also behave chaotically. Consider the pure ⁻feedback operator

$$X_t = 1 - \Lambda X_{t-1}, \tag{R21.2}$$

which has an equilibrium point $X_t = X_{t-1}$ at $1/(1 + \Lambda)$. If $\Lambda < 1$, then oscillations about the equilibrium point die out while if $\Lambda > 1$ the oscillations grow in amplitude. Hence, if x_0 is an error of estimation of the true population density, then the expected value of X_t will be

$$E[X_t] = 1 - \Lambda(X_{t-1} - x_{t-1}). \tag{R21.3}$$

Equations (R21.2) and (R21.3) have the solutions

$$X_t = 1 - \Lambda + \Lambda^2 + \Lambda^3 + \ldots \Lambda^{t-1} + \Lambda^t X_0, \qquad (R21.4)$$

$$E[X_t] = 1 - \Lambda + \Lambda^2 + \Lambda^3 + \ldots \Lambda^{t-1} + \Lambda^t (X_0 - x_0), \qquad (R21.5)$$

from which

$$x_t = X_t - E[X_t] = x_0 \Lambda^t = e^{\lambda x_0}, \quad \lambda = \ln\Lambda, \qquad (R21.6)$$

and, therefore, equation (R21.6) is identical to equation (R21.1). Once again, chaotic dynamics occur when $\lambda > 0$.

The search for chaos in nature

The search for complex aperiodic dynamics exhibited by certain mathematical models like the discrete logistic equation (Chapter 5) has been a popular activity amongst ecologists in the last decade or so (see, e.g. Logan and Hain (1991)). However, the early claims that chaos is common in ecology (Schaffer and Kot, 1986) have not generally held up (see, e.g. Turchin and Taylor (1992) and a counter argument by Perry *et al.* (1993)) and, in fact, there may be good reasons why ecological systems should not be expected to be chaotic (Berryman and Millstein, 1989). One case where chaotic oscillations may be evident is in populations of boreal voles (Hanski *et al.*, 1993).

Berryman, A. A. and Millstein, J. A. (1989) Are ecological systems chaotic – and if not, why not? *Trends in Ecology and Evolution* 4, 26–28.

Hanski, I., Turchin, P., Korpimaki, E. and Henttonen, H. (1993) Population oscillations of boreal rodents: regulation by mustelid predators leads to chaos. *Nature* 364, 232–235.

Logan, J. A. and Hain, F. P. (1991) *Chaos and Insect Ecology*. Virginia Experiment Station Information Series 91–3. Virginia Polytechnic Institute and State University, Blacksburg.

Maruyama, M. (1968) Mutual causality in general systems, in: J. H. Milsum (Ed.) *Positive Feedback: a General Systems Approach to Positive/Negative Feedback and Mutual Causality*. Pergamon Press, Oxford.

Perry, J. N., Woiwod, I. P. and Hanski, I. (1993) Using response-surface methodology to detect chaos in ecological time series. *Oikos* 68, 329–339.

Schaffer, W. M. and Kot, M. (1986) Chaos in ecological systems: the coals that Newcastle forgot. *Trends in Ecology and Evolution* 1, 58–63.

Turchin, P. and Taylor, A. D. (1992) Complex dynamics in ecological time series. *Ecology* 73, 289–305.

22. Circular causal pathways

Hutchinson (1948) was one of the first to recognize the importance of circular causal pathways in ecology and their potential effects on population dynamics. It is now widely recognized that mutual or circular causality can generate extremely complex patterns in time and space (Maruyama, 1968)[21].

Hutchinson, G. E. (1948) Circular causal systems in ecology. *Proceedings of the New York Academy of Sciences* 50, 221–246.

23. Ratio-dependent predation

There are two major philosophies in modelling predator–prey interactions:

1. The classical approach appeals to the physical "law of mass action" describing the rate of collision between molecules in a randomly mixed fluid or gas. From this law springs the so-called *Lotka–Volterra* models of predator–prey interactions and food web dynamics. Lotka–Volterra models are sometimes called "prey-dependent" because predator feeding rates (behavioural responses[10]) and per capita rates of change are only affected by prey density.
2. The alternative approach appeals to the economic law of "diminishing returns", as we do in this book. Models arising from the application of this principle are *ratio-dependent* because predator feeding rates and per capita rates of change are determined by the predator/prey (or prey/predator) ratio.

These two views of trophic interactions make quite different predictions about the dynamics and steady-state behaviour of ecosystems composed of many interacting species, some of which are discussed by Akçakaya *et al.* (1995), Arditi and Ginzburg (1989), and Michalski and Arditi (1995). Arguments for and against these two approaches can be found in Arditi *et al.* (1991), Abrams (1994), Gleeson (1994), Sarnelle (1994), Berryman *et al.* (1995), and Akçakaya *et al.* (1995). The logistic-type equations used in this book are one kind of ratio-dependent model. Other authors use ratio-dependent modifications of the classical Lotka–Volterra equations (e.g. Beddington *et al.*, 1975; DeAngelis *et al.*, 1975; Arditi and Ginzburg, 1989), or physiologically based ratio-dependent models (Gutierrez, 1992) based on Ivlev–Watt functional responses[10].

Abrams, P. A. (1994) The fallacies of ratio-dependent predation. *Ecology* 75, 1842–1850.

Akçakaya, H. R., Arditi, R. and Ginzburg, L. R. (1995) Ratio-dependent predation: an abstraction that works. *Ecology* 76, 995–1004.

Arditi, R. and Ginzburg, L. R. (1989) Coupling in predator–prey dynamics: ratio-dependence. *Journal of Theoretical Biology* 139, 311–326.

Arditi, R., Ginzburg, L. R. and Akcakaya, H. R. (1991) Variations in plankton densities among lakes: a case for ratio-dependent models. *American Naturalist* 138, 1287–1296.

Beddington, J. R., Free, C. A. and Lawton, J. H. (1975) Dynamic complexity in predator–prey models framed in difference equations. *Nature* 225, 58–60.

Berryman, A. A., Gutierrez, A. P. and Arditi, R. (1995) Credible, parsimonious, and useful predator–prey models: a reply to Abrams, Gleeson, and Sarnelle. *Ecology* 76, 1980–1985.

DeAngelis, D. L., Goldstein, R. A. and O'Neill, R. V. (1975) A model for trophic interactions. *Ecology* 56, 881–892.

Gleeson, S. K. (1994) Density dependence is better than ratio dependence. *Ecology* 75, 1834–1835.

Gutierrez, A. P. (1992) Physiological basis of ratio-dependent predator–prey theory: the metabolic pool model as a paradigm. *Ecology* 73, 1552–1563.

Michalski, J. and Arditi, R. (1995) Food web structure at equilibrium and far from it: is it the same? *Proceedings of the Royal Society of London* B, 259, 217–222.

Sarnelle, O. (1994) Inferring process from pattern: trophic level abundances and imbedded interactions. *Ecology* 75, 1835–1841.

24. The order of dynamic processes and time delays

The order of a dynamic process is defined following Royama (1977). Let X_t be the value of a variable at time t, then a *first-order* dynamic process is defined by the relationship

$$X_t = g(X_{t-1}).$$

Notice that the feedback in this system only involves one variable. The dynamics of such processes are characterized by sharp, high-frequency, or saw-toothed, oscillations known as *first-order dynamics*.

Now suppose that X interacts mutually with another variable Y, then we can write a system of two first-order difference equations

$$X_t = g(X_{t-1}, Y_{t-1})$$
$$Y_t = h(Y_{t-1}, X_{t-1})$$

that can be reduced to a *second-order* difference equation in either variable as follows: first invert the first equation with respect to Y_{t-1}

$$Y_{t-1} = g^*(X_t, X_{t-1}),$$

where g^* indicates the inversion of $g(X,Y)$ with respect to Y. [Note that this assumes the inversion has only one solution, which seems reasonable for most ecological systems (see Royama, 1977)]. Then substitute this inversion for Y_{t-1} in the second equation

$$Y_t = h[X_{t-1}, g^*(X_t, X_{t-1})],$$

and substitute this equation for Y_{t-1} in the first equation, noting that t has to be decremented by one ($t = t - 1, t - 1 = t - 2$), then

$$X_t = g\{X_{t-1}, h[X_{t-2}, g^*(X_{t-1}, X_{t-2})]\}.$$

This equation can also be written in more general terms

$$X_t = G(X_{t-1}, X_{t-2}).$$

This equation describes a *second-order* process because the state of X is determined by its state in two previous time periods. Notice that a completely connected system of two first-order processes can be reduced to a single second-order process in one variable, and that transformation manifests itself as a *time delay* in the feedback response. With some caution we can generalize this to a system composed of d interacting variables

$$X_t = G(X_{t-1}, X_{t-2}, \ldots X_{t-d}),$$

realizing that a solution to the inversion becomes more tenuous as the dimension of the system d increases.

Royama, T. (1977) Population persistence and density dependence. *Ecological Monographs* 47, 1–35.

25. Feedback dominance and limiting factors

The "law of the minimum" was formulated empirically by experimentally manipulating plant nutrients (Liebig, 1840). The results showed that, although plant growth depends on many nutrients, it is the one in shortest supply that actually limits growth. This law was later restated as the "law of limiting factors" (Blackman, 1905) and has been integrated into the concept of feedback dominance by Berryman (1993). Support for these ideas can be found in the empirical studies of Paine (1980, 1992) and the theoretical work of Michalski and Arditi (1995)[23]. Both studies suggest that many (or most) of the potential interactions in complex food webs become non-functional as the system evolves towards equilibrium, and that community dynamics become dominated by a few *strong interactions* or dominant feedbacks. The strong interactions are often associated with a predator at the top of the food chain that utilizes, and regulates the abundance of, many of the other species in the community, the so-called "keystone predator" (Paine, 1966).

Berryman, A. A. (1993) Food web connectance and feedback dominance, or does everything really depend on everything else? *Oikos* 68, 183–185.

Blackman, F. F. (1905) Optima and limiting factors. *Annals of Botany* 19, 281–295.

Liebig, J. (1840) *Chemistry and its Application to Agriculture and Physiology.* Taylor and Walton, London.

Paine, R. T. (1966) Food web complexity and species diversity. *American Naturalist* 100, 65–75.

Paine, R. T. (1980) Food webs: linkage, interaction strength and community infrastructure. *Journal of Animal Ecology* 49, 667–685.

Paine, R. T. (1992) Food-web analysis through field measurement of per-capita interaction strength. *Nature* 355, 73–75.

26. Limiting groups and hierarchies

An example of aphid populations being limited by a group of insect predators is given by Morris (1992), and the concept of forest insect populations being regulated by a hierarchy of different groups of limiting factors was developed by Berryman *et al.* (1987).

Berryman, A. A., Stenseth, N. C. and Isaev, A. S. (1987) Natural regulation of herbivorous forest insect populations. *Oecologia* 71, 174–184.

Morris, W. F. (1992) The effects of natural enemies, competition, and host plant water availability on an aphid population, *Oecologia* 90, 359–365.

27. Genetic and physiological time delays

If population density can induce genetic and/or physiological changes in offspring of the next generation, then time delays can be introduced into the feedback structure and this can cause dynamic patterns similar to those created by interactions with reactive environmental factors. For example, at high population densities the amount of yolk provided to eggs could be lower because females have lower energy reserves (less available food), and this could reduce the survival of offspring in the next generation, the so-called *maternal effect* (Rossiter, 1994; Ginzburg and Tanneyhill, 1994; but see also Berryman (1995)). Another possibility is that certain *genotypes* could survive better at high densities (e.g. less fecund or more territorially aggressive genotypes) and this could lead to lower reproduction in the following generations (Chitty, 1967; Witting, 1997; but see Stenseth (1981)). In a similar way, time delays can be introduced into feedbacks involving competition for passive resources if the *quality* of the resource is affected by past population densities. Examples can be found in the feeding of caterpillars on deciduous trees which induces changes in the quality of foliage produced in the following years (Baltensweiler and Fischlin, 1988; Haukioja *et al.*, 1988).

Baltensweiler, W. and Fischlin, A. (1988) The larch budmoth in the Alps, in: A. A. Berryman (Ed.) *Dynamics of Forest Insect Populations: Patterns, Causes, Implications*, pp. 331–351. Plenum Press, New York.

Berryman, A. A. (1995) Population cycles: a critique of the maternal and allometric hypotheses. *Journal of Animal Ecology* 64, 290–293.

Chitty, D. (1967) The natural selection of self-regulatory behavior in animal populations. *Proceedings of the Ecological Society of Australia* 2, 51–78.

Ginzburg, L. R. and Taneyhill, D. E. (1994) Population cycles of forest Lepidoptera: a maternal effect hypothesis. *Journal of Animal Ecology* 63, 79–92.

Haukioja, E., Neuvonen, S., Hanhimaki, S. and Niemela, P. (1988) The autumnal moth in Fennoscandia, in: A. A. Berryman (Ed.) *Dynamics of Forest Insect Populations: Patterns, Causes, Implications*, pp. 163–178. Plenum Press, New York.

Rossiter, M. C. (1994) Maternal effects hypothesis of herbivore outbreak. *BioScience* 44, 752–763.

Stenseth, N. C. (1981) On Chitty's theory for fluctuating populations: the importance of genetic polymorphism in the generation of regular density cycles. *Journal of Theoretical Biology* 90, 9–36.

Witting, L. (1997) *A General Theory of Evolution by Means of Selection by Density Dependent Competitive Interactions*. Peregrine, Arhus.

28. Meta-stability

The term meta-stability was originally used to describe physical systems with apparently stable equilibrium points that suddenly undergo quantum transitions following minor exogenous disturbances. For example, water under pressure can be heated to several degrees above its boiling point without boiling, but a small external disturbance will trigger its sudden transition to the vapour state. Lotka (1925) seems to have been the first to use this term to describe ecological systems with bounded stability domains. In this sense, the ecological system remains in a stable state (the point J in Figure 7.3) until it is disturbed beyond an unstable boundary (the point E in Figure 7.3), after which it undergoes a sudden transition to a new equilibrium (the point K), or exhibits catastrophic "boom and bust" dynamics (Figure 7.6, bottom and see Figure 29 in Lotka's book).

Berryman *et al.* (1984) provide an example of meta-stable dynamics in forest ecosystems infested by bark beetles (see also Chapter 14).

Berryman, A. A., Stenseth, N. C. and Wollkind, D. J. (1984) Metastability in forest ecosystems infested by bark beetles. *Researches in Population Ecology* 26, 13–29.
Lotka, A. J. (1925) *Elements of Physical Biology.* Republished in 1956 by Dover, New York.

29. Classification of insect outbreaks

The classification scheme developed in this book is the same as that proposed by Berryman (1978, 1987, 1990). This scheme is very similar to that developed independently by Russian scientists (Isaev *et al.*, 1984). Both these systems were derived from the fundamental theory of population dynamics but the terminology is somewhat different. Nedorezov (1997) has recently summarized the Russian approach in the English language. Other classification schemes are less general in that they are only meant to apply to particular groups, such as defoliating insects.

Berryman, A. A. (1978) Towards a theory of insect epidemiology. *Researches in Population Ecology* 19, 181–196.
Berryman, A. A. (1987) The theory and classification of outbreaks, in: P. Barbosa and J. C. Schultz (Eds) *Insect Outbreaks*, pp. 3–30. Academic Press, New York.
Berryman, A. A. (1990) Identification of outbreak classes. *Mathematical and Computer Modelling* 13, 105–116.
Isaev, A. S., Khlebopros, R. G., Nedorezov, L. V., Kondakov, Y. P. and Kiselev, V. V. (1984) *Population Dynamics of Forest Insects* (in Russian). Nauka Publishing House, Moscow.
Nedorezov, L. V. (1997) Escape effects and population outbreaks. *Ecological Modelling* 94, 95–110.

30. Characteristics of outbreak species

One of the first attempts to identify the special attributes of outbreak species was by Southwood and Commins (1976) who argued that r-selected species with their high reproductive and dispersal powers should be more likely to exhibit outbreaks than K-selected species (see also Stenseth (1987)). Note that r-selected species are often adapted to utilizing rare or ephemeral habitats and outbreaks occur when such habitats become abundant because of environmental changes. Many have noted the tendency for outbreak insect species to lay their eggs in groups and for their larvae to feed in aggregations (e.g. Wallner, 1987; Hunter, 1991). Others see the lack of selection of oviposition sites by females (Price, 1994), the tendency for reproduction to be dependent on larval rather than adult feeding (capital breeders; Tammaru and Haukioja, 1996), and polymorphism (Wallner,

1987) as important characteristics of outbreak species. In addition, insects that exhibit outbreaks in hardwood forests tend to hibernate in the egg stage so that they begin feeding early in spring when new foliage first becomes available (Hunter, 1991). Hunter also found that outbreak species were generally poorer fliers (in contrast to Southwood and Commins (1976)) and had a more varied diet.

Hunter, A. F. (1991) Traits that distinguish outbreaking and nonoutbreaking Macrolepidoptera feeding on northern hardwood trees. *Oikos* 60, 275–282.

Price, P. W. (1994) Phylogenetic constraints, adaptive syndromes, and emergent properties: from individuals to population dynamics. *Researches in Population Dynamics* 36, 3–14.

Southwood, T. R. E. and Commins, H. N. (1976) A synoptic population model. *Journal of Animal Ecology* 45, 949–965.

Stenseth, N. C. (1987) Evolutionary processes and insect outbreaks, in: P. Barbosa and J. C. Schultz (Eds) *Insect Outbreaks*, pp. 533–563. Academic Press, New York.

Tammaru, T. and Haukioja, E. (1996) Capital breeders and income breeders among Lepidoptera – consequences to population dynamics. *Oikos* 77, 561–564.

Wallner, W. E. (1987) Factors affecting insect population dynamics: differences between outbreak and non-outbreak species. *Annual Review of Entomology* 32, 317–340.

31. Cowpea weevil data

Utida (1967) reared southern cowpea weevils, *Callosobruchus maculatus*, on a fixed quantity of food (10 grams or 50–60 azuki beans supplied to each generation). Adult weevils were counted at the end of each generation. The experiment ran for 10 weevil generations. The time series shown in Figure 9.1 was estimated directly from Utida's experiment A in his Figure 1.

Utida, S. (1967) Damped oscillation of population density at equilibrium. *Researches in Population Ecology* 9, 1–9.

32. Statistics

The **mean**, $\bar{X}$, is the average value of a series of numbers. For example, if we have the numbers 4, 22, 15, 9, then the mean of the series is $4 + 22 + 15 + 9 = 50/4 = 12.5$. The general formula for the mean is, therefore,

$$\bar{X} = \frac{\sum_{i=1}^{I} X_i}{I}$$

where X_i is the value of the ith variable in a string of length I and $\sum_{i=1}^{I} X_i$ means the sum of all X_i, $i = 1,2,3, \ldots I$.

The **variance**, s^2, is a measure of the degree of variation in a series of numbers around its mean value. The general formula for the variance of a series is

$$s^2 = \frac{\sum_{i=1}^{I}(X_i - \overline{X})^2}{I - 1}$$

or, alternatively,

$$s^2 = \frac{I\sum_{i=1}^{I} X_i^2 - \left(\sum_{i=1}^{I} X_i\right)^2}{I - 1}$$

The values are squared to make all quantities larger than zero. Obviously if all the numbers in a series are the same, the variance will be zero, while if the numbers are very different the variance will be large.

The **standard deviation**, s, is the square root of the variance.

Least squares regression. Models are commonly fitted to data with *least squares regression*, a method for finding the best fit of a model by minimizing the sum of squares of deviations of the data from the regression line. In regression analysis we have a dependent variable on the Y-axis which is associated with an independent variable on the X-axis. If this association is linear, then the appropriate model is $Y = a + bX$.

In "least squares" regression, the best fit is found by minimizing the sum of squares, SSQ, of the differences between the observed values of the dependent variable, Y_i, and the expected values predicted by the model, $E[Y_i]$; i.e.

$$SSQ = \sum_{i=1}^{I}(Y_i - E[Y_i])^2 \,.$$

Linear models can be fitted by solving for the slope of the regression

$$b = \frac{I\sum_{i=1}^{I} X_i Y_i - \left(\sum_{i=1}^{I} X_i\right)\left(\sum_{i=1}^{I} Y_i\right)}{I\sum_{i=1}^{I} X_i^2 - \left(\sum_{i=1}^{I} X_i\right)} \,.$$

The intercept of the regression line with the Y-axis is then given by

$$a = \overline{Y} - b\overline{X}$$

where a is the intercept and the overbar identifies the means of the two series of observations.

Non-linear convergence. Non-linear models can be fitted to data with *convergence* algorithms. In this case the values of the parameters are changed in an organized manner until the sum of squares is minimized. The convergence algorithm is usually started with the parameter values obtained by linear regression.

Variation about the regression line. In least squares regression we find the best fit to the data by minimizing the deviations between data and model. After this has been done, the residual variation of data from the fitted model is assumed to be due to random variability. If Y_i is the actual value of the dependent variable associated with the value X_i of the independent variable, then the expected value of Y_i according to the model is $E[Y_i] = a + bX_i$, and the variance of the observed value from this expectation is $Y_i - (a + bX_i)$. The variance about the regression is, therefore,

$$s^2 = \frac{\sum_{i=1}^{I}[Y_i - (a + bX_i)]^2}{I - 2}$$

or, alternatively,

$$s^2 = \frac{\sum_{i=1}^{I}Y_i^2 - a\sum_{i=1}^{I}Y_i - b\sum_{i=1}^{I}X_iY_i}{I - 2}.$$

Note that there are two degrees of freedom because two parameters a and b are used in the calculation. The variation about a regression line is sometimes called the *error of variance* or the *standard error of regression*. Naturally the *standard deviation*, s, about the regression is the square root of the variance.

Correlation. The *correlation coefficient*, r, measures the degree of association between the two variables given a particular theoretical model. Statistical correlation for the linear model is computed from

$$r = \frac{I\sum_{i=1}^{I}X_iY_i - \left(\sum_{i=1}^{I}X_i\right)\left(\sum_{i=1}^{I}Y_i\right)}{\sqrt{\left[I\sum_{i=1}^{I}X_i^2 - \left(\sum_{i=1}^{I}X_i\right)^2\right]\left[I\sum_{i=1}^{I}Y_i^2 - \left(\sum_{i=1}^{I}Y_i\right)^2\right]}}.$$

A correlation coefficient of 1 means that the relationship between the two variables is perfect, while a value of zero means that the two variables are completely unrelated.

The "goodness of fit" of a model is measured by the *coefficient of determination*, r^2. This is sometimes multiplied by 100 to give the percentage of the variation in the data explained by the model. In the cowpea weevil example, an r^2 of 0.99 means that 99% of the variation in R is explained by the non-linear R-function, so that only 1% of the variability in R is due to exogenous random environmental variability.

Autocorrelation. Given a time series N_t, $t = 1, 2, 3, \ldots I$, then the autocorrelation at lag d is the correlation coefficient, r_d, calculated by setting $Y_i = \ln N_t$ and $X_i = \ln N_{t-d}$, $i = 1, 2, 3 \ldots I - d$.

Partial autocorrelation. The partial autocorrelation PA_d gives the correlation between $Y_i = \ln N_t$ and $X_i = \ln N_{t-d}$ with the effects of lesser lags removed. It can be calculated by first computing the autocorrelation at each lag, r_d, $d = 1, 2, 3 \ldots$ and then sequentially calculating PA_d as follows

$$PA_1 = r_1$$

$$PA_d = \frac{r_d - \sum_{j=1}^{d-1} PA_{d-1,j} r_{d-j}}{1 - \sum_{j=1}^{d-1} PA_{d-1,j} r_j}$$

where

$$PA_{d,j} = PA_{d-1,j} - PA_{d,j} PA_{d-1,j-1}$$

and

$$j = 1, 2, 3, \ldots d - 1.$$

Partial rate correlation. The partial rate correlation at lag d, PR_d, can be calculated as follows: first calculate the correlation coefficient between $Y_i = R = \ln N_t - \ln N_{t-1}$ and $X_i = \ln N_{t-1}$ for the $I - 1$ data points in the series. This is the partial rate correlation at lag 1, PR_1. Next, calculate the partial autocorrelation for lags 2 and higher with the formula above. The partial rate correlation at lags greater than 1 are identical to the partial autocorrelation at the same lags; i.e. $PR_d = PA_d$, $d > 1$.

33. Biological control of the gypsy moth

A controversy has arisen over the role of parasitoids in the biological control of gypsy moth (Montgomery and Wallner, 1988; Berryman, 1991a, b; Liebhold and Elkinton, 1991). This disagreement revolves around an experiment by Gould *et al.* (1990) in which they transplanted gypsy moth egg masses from dense populations into sparse populations[14]. Test populations were almost eliminated by insect parasitoids, supporting the hypothesis that parasitoids are the limiting factor for sparse gypsy moth populations. Because these parasitoids are generalists and, therefore, require other host species to complete their annual cycle, it is possible for gypsy moth populations to escape control and rise to outbreak densities (second principle). On the other hand, there is also evidence that sparse gypsy moth populations are limited by small mammals, and that escape to outbreak densities occurs when the behavioural responses of these predators are saturated by high gypsy moth numbers (Campbell and Sloan, 1977, 1978).

Berryman, A. A. (1991a) The gypsy moth in North America: a case of successful biological control? *Trends in Ecology and Evolution* 6, 110–111.

Berryman, A. A. (1991b) Reply from Alan Berryman. *Trends in Ecology and Evolution* 6, 264.

Campbell, R. W. and Sloan, R. J. (1977) Natural regulation of innocuous gypsy moth populations. *Environmental Entomology* 6, 315–322.

Campbell, R. W. and Sloan, R. J. (1978) Numerical bimodality among North American gypsy moth populations. *Environmental Entomology* 7, 641–646.

Liebhold, A. M. and Elkinton, J. S. (1991) Gypsy moth dynamics. *Trends in Ecology and Evolution* 6, 263–264.

Montgomery, M. E. and Wallner, W. E. (1988) The gypsy moth: a westward migrant, in A. A. Berryman (Ed.) *Dynamics of Forest Insect Populations.* Plenum, New York.

34. The fir engraver beetle

The analysis and interpretation of population changes in the fir engraver beetle can be found in Berryman (1973) and Berryman and Ferrell (1988). Berryman and Pienaar (1973) presented a detailed analysis of the effects of *intra*specific competition for food on the survival of this bark beetle. These analyses led to the conclusion that the endogenous dynamics of the fir engraver population are largely driven by first-order *intra*specific competition for food and that any higher-order cyclic tendencies are the result of exogenous factors.

Berryman, A. A. (1973) Population dynamics of the fir engraver beetle, *Scolytus ventralis* (Coleoptera: Scolytidae). I. Analysis of population behavior and survival from 1964–1971. *Canadian Entomologist* 105, 1465–1488.
Berryman, A. A. and Ferrell, G. T. (1988) The fir engraver beetle in western States, in: A. A. Berryman (Ed.) *Dynamics of Forest Insect Populations.* Plenum Press, New York.
Berryman, A. A. and Piennar, L. V. (1973) Simulation of intraspecific competition and survival of *Scolytus ventralis* broods (Coleoptera: Scolytidae). I. Analysis of population behavior and survival from 1964–1971. *Environmental Entomology* 2, 447–459.

35. Hypothesis testing with models

A good example of using models to test hypotheses about the causes of population fluctuations can be found in the work of Hanski and his associates on vole populations (Hanski *et al.*, 1991, 1993; Hanski and Korpimaki, 1995; Turchin and Hanski, 1997). Diagnostic analyses of time series data collected throughout Fennoscandia indicated that vole populations exhibited first-order dynamics in the south and higher-order cycles in the north. One hypothesis for this change in the order of the dynamics was that vole populations were limited by the switching/aggregation of non-reactive generalist predators in the south, where such predators are abundant, and by interaction with reactive specialist predators in the north, where generalists are much more rare. A model of this system was then built and the parameters estimated from independent data. Simulations with the model produced dynamic patterns very similar to those observed in nature, and this was taken as support for the hypothesis.

Hanski, I., Hansson, L. and Henttonen, H. (1991) Specialist predators, generalist predators, and the microtine rodent cycle. *Journal of Animal Ecology* 60, 353–367.
Hanski, I. and Korpimaki, E. (1995) Microtine rodent dynamics in northern Europe: parameterized models for the predator–prey interaction. *Ecology* 76, 840–850.

Hanski, I., Turchin, P., Korpimaki, E. and Henttonen, H. (1993) Population oscillations of boreal rodents: regulation by mustelid predators leads to chaos. *Nature (London)* 364, 232–235.

Turchin, P. and Hanski, I. (1997) An empirically based model for latitudinal gradient in vole population dynamics. *American Naturalist* 149, 842–874.

36. Soay sheep data

Soay sheep, a primitive domesticated sheep that closely resembles the original wild species, were introduced into Hirta in 1932, and quickly increased to the carrying capacity of the island. Sheep numbers were counted from 1961 to 1967 by Jewell *et al.* (1974) and from 1985 to 1990 by Clutton-Brock *et al.* (1991).

Clutton-Brock, T. H., Price, O. F., Albon, S. D. and Jewell, P. A. (1991) Persistent instability and population regulation in Soay sheep. *Journal of Animal Ecology* 60, 593–608.

Jewell, P. A., Miller, C. and Boyd, J. M. (1974) *Island Survivors: the Ecology of the Soay Sheep of St Kilda*. Athlone Press, London.

37. Bias in regression analysis

When the realized per capita rate of change is calculated from the relationship $R = \ln N_t - \ln N_{t-1}$ and R is used as a dependent variable in the regression of R on N_{t-1}, then the independent variable, N_{t-1}, is not truly independent because it appears on both sides of the regression equation. This introduces bias into the estimation of the parameters of the R-function, particularly at lag 1, but less so for higher lags. Although this problem should not distract us from building R-function models from time series data, we should be aware that it exists and that the high correlations obtained when fitting lag 1 models to time series data may be partly due to this bias.

38. Multiple delay models

The Gompertz equation with multiple lags has been used extensively by Royama[16, 24] while more complicated *response surface*, *neural net* and *spline* methods have been employed by Ellner and Turchin (1995) to model logistic R-functions.

Ellner, S. and Turchin, P. (1995) Chaos in a noisy world: new methods and evidence from time series analysis. *American Naturalist* 145, 343–375.

39. Forecasting human population growth

Two elementary models have been proposed for the growth of the human population, both of which were developed from differential equations, which may be more appropriate for populations of organisms like humans that breed continuously and have completely overlapping generations. The first is an empirical equation developed by von Foerster *et al.* (1960)

$$N_t = \frac{1.79 \times 10^{11}}{(2027 - t)^{0.99}},$$

where N_t is the number of people in year t. This model, published in 1960, accurately predicts the number of humans on the planet 35 years later (Berryman and Valenti, 1994); for example, it predicts a population of 5791 million by 1995 compared with the United Nation's figure of 5733 million in 1995.

The second model (Dochy, 1995)

$$N_t = \frac{1}{9.833 \times 10^{-3} - (4.849 \times 10^{-6})t},$$

overestimates the population in 1995 at 6280 million. Both the von Foerster and Dochy models predict that the human population will reach an impossible *infinite* growth rate by about the year 2027, called "doomsday" by von Forester *et al.* (1960). The data used in this book were taken from Dochy's paper.

More detailed models of human population dynamics have been developed by the "Club of Rome Project on the Predicament of Mankind" (Meadows *et al.*, 1972, 1992). These models include the interaction of the human population with natural resources, food production, pollution and industrial output, and generally predict the cessation of runaway population growth, and sometimes dramatic collapses of industrial system and population, by the middle of the 21st century.

Berryman, A. A. and Valenti, M. A. (1994) The doomsday prediction. *Bulletin of the Ecological Society of America* 75, 123–124.

Dochy, F. (1995) Human population growth: local dynamics-global effects. *Acta Biotheoretica* 43, 241–247.

Meadows, D. H., Meadows, D. L. and Randers, J. (1992) *Beyond the Limits: Confronting Global Collapse, Envisioning a Sustainable Future*. Chelsea Green Publishing Company, Post Mills, Vermont.

Meadows, D. H., Meadows, D. L., Randers, J. and Behrens, W. W. (1972) *The Limits to Growth*. Universe Books, New York.

Von Foerster, P., More, M. and Amiot, L. W. (1960) Doomsday: Friday, 13 November, A. D. 2026. *Science* 132, 1291–1295.

40. Economic threshold

The concept of economic threshold (ET) is complicated by the problem of giving a value to human health and well-being, from both economic and aesthetic perspectives; i.e. what is the value of a mountain scene, a burbling brook? The ET usually occurs at some pest density below the *economic injury level* or EIL, the density of pests where the economic losses from the pest are equal to the cost of control, and is therefore the pest density at which control is necessary to prevent economic damage. For more information on these concepts of pest control, see:

Horn, D. J. (1988) *Ecological Approaches to Pest Management*. Guilford Press, New York.

Pedigo, L. P. (1996) *Entomology and Pest Management*. Prentice Hall, New Jersey.

41. Sandhill crane

The spring sandhill crane census from 1959–1978 were (numbers divided by 100): 1475, 1259, 1363, 1428, 1019, 1560, 803, 1231, 1230, 1692, 1550, 1936, 2075, 1836, 1953, 1770, 2275, 1538, 2191, 1599 (Buller, 1979). Estimates of hunting mortality were obtained from Johnson (1979) who also developed a simulation model for crane population dynamics.

Buller, R. J. (1979) Lesser and Canadian sandhill crane populations, age struc-
    ture, and harvest. United States Department of the Interior, Fish and Wildlife
    Service, Special Scientific Report, Wildlife No. 221.
Johnson, D. H. (1979) Lesser and Canadian sandhill crane populations, age struc-
    ture, and harvest. United States Department of the Interior, Fish and Wildlife
    Service, Special Scientific Report, Wildlife No. 221.

42. Cone beetle

Information on the population dynamics of cone beetles and red pine cones was obtained from Mattson (1980). The data used in the analysis are presented in Table R1.

Mattson, W. J. (1980) Cone resources and the ecology of the red pine cone beetle,
    Conophthorus resinosae (Coleoptera: Scolytidae). *Annals of the Entomological
    Society of America* 73, 390–396.

Table R1 Data from Mattson (1980) on the abundance of red pine cones and cone beetles per 0.1 acres in a Wisconsin seed production area (original data were presented in numbers per acre but are divided by 10 for our analysis)

Year	Cones	Attacked cones	Beetles
1969	2363.2	1725.14	121.8
1970	1463.6	1463.60	720.0
1971	1300.6	1026.17	152.5
1972	2861.5	2520.98	271.8
1973	2003.9	1803.51	177.0
1974	1875.5	1781.72	161.4
1975	3541.9	2302.23	105.5
1976	718.9	718.90	369.5
1977	1413.1	712.20	50.0
1978	2154.4	2111.31	100.0

43. Spruce needleminer

Data on the spruce needleminer were collected over a 19-year period in a spruce stand (Stand A) in Zealand, Denmark (Munster-Swendsen (1985) and personal communication) (Table R2). Additional information on the parasitoids and their role in the population dynamics of the spruce needleminer can be found in Munster-Swendsen (1979, 1980, 1982, 1994).

Munster-Swendsen, M. (1979) The parasitoid complex of *Epinotia tedella* (Cl.) (Lepidoptera: Tortricidae). *Entomologiske Meddelelser* 47, 63–71.

Munster-Swendsen, M. (1980) The distribution in time and space of parasitism in *Epinotia tedella* (Cl.) (Tortricidae). *Ecological Entomology* 5, 373–383.

Munster-Swendsen, M. (1982) Interactions within a one-host–two-parasitoids system, studied by simulation of spatial patterning. *Journal of Animal Ecology* 51, 97–110.

Munster-Swendsen, M. (1985) A simulation study of primary-, clepto- and hyper-parasitism in *Epinotia tedella* (Cl.) (Lepidoptera: Tortricidae). *Journal of Animal Ecology* 54, 683–695.

Munster-Swendsen, M. (1994) Pseudoparasitism: detection and ecological significance in *Epinotia tedella* (Cl.) (Tortricidae). *Norwegian Journal of Agricultural Sciences* Supplement 16, 329–335.

Table R2 Data from Munster-Swendsen (1985 and personal communication) on the abundance of spruce needleminers descending per 10 square metres of forest floor and numbers containing insect parasitoids in a Norway spruce stand (stand A) in Zealand, Denmark (original data were in numbers per square metre but are multiplied by 10 for analysis)

Year	Needleminers per 10 square metres	Parasitoids per 10 square metres
1970	570.8	260.3
1971	2119.7	883.9
1972	3365.4	1771.2
1973	1006.4	517.4
1974	255.7	105.5
1975	695.7	205.2
1976	1055.4	454.9
1977	2420.0	1195.0
1978	2152.5	1677.4
1979	77.3	60.0
1980	13.4	7.4
1981	10.5	3.2
1982	132.4	25.6
1983	355.0	50.9
1984	1935.2	421.3
1985	6986.4	3060.7
1986	5594.7	4004.7
1987	767.9	412.0
1988	127.8	83.1

44. Mountain pine beetle

Data for an endemic mountain pine beetle population on Starvation Ridge (Figure 14.1, left) came from Tunnock (1970) and the epidemic population in Yellowstone Park (Figure 14.1, right) from Parker (1973). Figure 14.2 was reproduced from McGregor *et al.* (1983). The role of host plant resistance in the population dynamics of the mountain pine beetle is described in Raffa and Berryman (1983). Explanations

and interpretations of the population dynamics of the mountain pine beetle follow
Berryman (1976, 1982a, b), and models of the interaction between beetles and pine
trees can be found in Berryman *et al.* (1984[28], 1989) and Raffa and Berryman (1986).
Lodgepole pine yield tables were obtained from Cole and Edminster (1985).

Berryman, A. A. (1976) Theoretical explanation of mountain pine beetle dynamics
in lodgepole pine forests. *Environmental Entomology* 5, 1225–1233.
Berryman, A. A. (1982a) Mountain pine beetle outbreaks in Rocky Mountain
lodgepole pine forests. *Journal of Forestry* 80, 410–413.
Berryman, A. A. (1982b) Population dynamics of bark beetles, in: J. B. Mitton
and K. B. Sturgeon (Eds) *Bark Beetles in North American Conifers – a System
for the Study of Evolutionary Ecology.* University of Texas Press, Austin.
Berryman, A. A., Raffa, K. F., Millstein, J. A. and Stenseth, N. C. (1989) Inter-
action dynamics of bark beetle aggregation and conifer defense rates. *Oikos* 56,
256–263.
Cole, D. M. and Edminster, C. B. (1985) Growth and yield of lodgepole pine, in:
D. M. Baumgatner, R. G. Krebil, J. T. Arnott and G. F. Weetman (Eds)
Lodgepole Pine: the Species and its Management. Cooperative Extension Service,
Washington State University, Pullman.
McGregor, M. D., Oakes, R. D. and Meyer, H. E. (1983) Status of mountain pine
beetle, Northern Region, 1982. *USDA Forest Service, Northern Region, State
and Private Forest Report* 83–16.
Parker, D. L. (1973) Trend of a mountain pine beetle outbreak. *Journal of Forestry*
71, 698–700.
Raffa, K. F. and Berryman, A. A. (1983) The role of host plant resistance in the
colonization behavior and ecology of bark beetles. *Ecological Monographs* 53,
27–49.
Raffa, K. F. and Berryman, A. A. (1986) A mechanistic computer model of mountain pine beetle populations interacting with lodgepole pine stands and its
implications for forest managers. *Forest Science* 32, 789–805.
Tunnock, S. (1970) A chronic infestation of mountain pine beetles in lodgepole
pine in Glacier National Park, Montana. *Journal of the Entomological Society
of British Columbia* 67, 23.

45. Rock lobsters

The interaction between rock lobsters and whelks is described by Barkai and
McQuaid (1988). Estimates of the equilibrium points K and M, the intercept of the
lobster isocline with the whelk axis, and the total area controlled by whelks around
Marcus Island, were obtained through private correspondence with Amos Barkai.

Barkai, A. and McQuaid, C. (1988) Predator–prey role reversal in a marine benthic
ecosystem. *Science* 242, 62–64.

46. The demographic transition

The term "demographic transition" refers to the decline in the human birth rate
that occurs in many industrializing societies, and the observation that reduced

fertility occurs earlier and more markedly in rich families. For a review of demographic transition theory from an evolutionary perspective, see Borgerhoff Mulder (1998), and a book edited by Coleman and Schofield (1986) is a good place to find details of modern demographic theory.

Borgerhoff Mulder, M. (1998) The demographic transition: are we any closer to an evolutionary explanation. *Trends in Ecology and Evolution* 13, 266–270.

Coleman, D. and Schofield, R. (1986) *The State of Population Theory: Forward from Malthus*. Basil Blackwell, Oxford.

47. The life table approach

The application of life tables to the study of insect population dynamics was pioneered in the 1950s by British and Canadian scientists working on the population dynamics of several important forest pests (e.g. Varley, 1947; Morris and Miller, 1954; Stark, 1958). Methods for analysing and modelling life table data were also developed around this time (e.g. Morris, 1963[15]; Varley *et al.*, 1973). A student-orientated presentation of the approach can be found in Begon and Mortimer (1986) and technical details in Southwood (1978)[6].

Begon, M. and Mortimer, M. (1986) *Population Ecology: a Unified Study of Animals and Plants*. Blackwell Scientific Publications, London.

Morris, R. F. and C. A. Miller (1954) The development of life tables for the spruce budworm. *Canadian Journal of Zoology* 32, 283–301.

Stark, R. W. (1958) Life tables for the lodgepole needleminer, *Recurvaria starki* Free. (Lepidoptera: Gelechiidae). *Proceedings of the 10th International Congress of Entomology* 4, 151–162.

Varley, G. C. (1947) The natural control of population balance in the Knapweed gall-fly (*Urophora jaceana*). *Journal of Animal Ecology* 16, 139–187.

Varley, G. C., Gradwell, G. R. and Hassell, M. P. (1973) *Insect Population Ecology: an Analytical Approach*. Blackwell Scientific Publications, London.

48. Time series analysis

The first attempts to use time series analysis to model ecological data seems to be the work of Moran (1953) on the long record of Canadian lynx fur returns maintained by the Hudson Bay Company. However, it was Tom Royama's monograph "Population persistence and density dependence"[24] that first presented a rigorous and logical framework for the analysis and interpretation of time series data from biological populations. Royama's approach is covered in detail in his book *Analytical Population Dynamics*[16]. This is the basic approach adopted by Turchin and others[21, 35, 38] as well as by myself. Although there are slight differences and idiosyncrasies between authors, they all have the same goal, the diagnosis, interpretation, modelling and prediction of population fluctuations.

Moran, P. A. P. (1953) The statistical analysis of the Canadian lynx cycle. I. Structure and prediction. *Australian Journal of Zoology* 1, 163–173.

Symbols

a The intercept of a regression line[32] with the ordinate (Y-axis) of the graph of Y on X.

A The *maximum* per capita rate of change of a particular species living in a given environment. This parameter represents the genetic potential of a species in that environment, or the difference between the maximum birth rate and the minimum death rate under the particular set of environmental and genetic conditions.

A_N The maximum per capita rate of change of species N, usually a prey species (see A).

A_P The maximum per capita rate of change of species P, usually a consumer or predator species (see A).

ACF   The autocorrelation[32] function. A histogram of the correlation between the logarithm of population density at time t and the logarithm of density at time $t - d$, where $d = 1, 2, 3, \ldots$

b The slope of a regression line[32].

B The per capita birth rate, or the number of births per individual in a population per unit of time.

c The quantity by which the survival and reproduction of the average organism in a population is reduced by the addition of a single competitor of the same species.

C The coefficient of *intra*specific competition (similar to c) for passive or non-reactive resources.

C_j The coefficient of *intra*specific competition (similar to c) for a low-density R-function (see J).

C_k The coefficient of *intra*specific competition (similar to c) for a high-density R-function (see K).

C_N The coefficient of *intra*specific competition (similar to c) for non-reactive resources within a population of species N, usually a prey species.

C_P The coefficient of *intra*specific competition (similar to c) for non-reactive resources within a population of species P, usually a predator or consumer species.

C_{NP} The coefficient of *intra*specific competition within a population of species N (a prey) when under attack by a population of predators P; i.e. the coefficient of *intra*specific competition for a reactive resource like enemy-free space. This parameter is directly related to W defined below.

C_{PN} The coefficient of *intra*specific competition amongst a population of species P for a reactive resource N, usually a prey species. This parameter is directly related to V defined below.

CAT   The control action threshold. The density of a pest at which a decision is made to initiate pest control action. Same as the economic threshold, ET.

CTS   The number of complete time steps that occur before a population attains its mean value following a disturbance.

d The time delay or time lag in the density induced ⁻feedback acting on the per capita rate of change, R.

d_j The time delay or time lag in the ⁻feedback operating in the vicinity of a low-density equilibrium, J.

d_k The time delay or time lag in the ⁻feedback operating in the vicinity of a high-density equilibrium, K.

D The per capita death rate. The probability that an average organism will die over a particular time interval, usually a year, when living in a particular environment.

D_Y The per capita death rate of young born in a particular time interval, usually a year.

D_A The per capita death rate of parental adults over a particular time interval, usually a year.

D_N The per capita death rate of a prey or resource species.

D_P The per capita death rate of a predator or consumer species.

e The base of the natural logarithm; $e = 2.71828 \ldots$

E The escape threshold. An *unstable equilibrium point* below which a population is regulated by predators or the defences of its prey, and above which it escapes to much higher densities.

EIL   The economic injury level or the density of pests where the economic losses from the pest are equal to the cost of control.

ET   The economic threshold. The density of a pest at which a decision is made to initiate pest control actions. Same as CAT.

E[.]   The expected value of the variable inside the brackets. Usually a prediction of the value of a variable made by a model.

f The behavioural or functional response of a consumer. The number of prey consumed by a predator in a unit of time expressed as a function of prey density.

$f(.)$ An undefined function of the arguments within the brackets.

F The relative abundance of alternative food for a consumer. This parameter represents the density of alternative prey available to the consumer, weighted by the relative preference for those prey over the central prey species, N.

$F(.)$ An undefined function of the arguments within the brackets.

G The finite per capita rate of change. The rate of change over a unit time period, usually a year, realized by an average organism living in a particular environment; i.e. $G = N_t/N_{t-1} = 1 + B - D$.

$g(.)$ An undefined function of the argument(s) within the brackets.

H The total amount of food (prey) available to a consumer (predator); i.e. $H = N + F$.

I The number of observations in a sample or time series.

J A low-density equilibrium maintained by generalist predators or prey resistance to attack; $J < K$, see below.

k The individual demand for a good (in economics) or a resource (in ecology). Similar, in the latter case, to the demand of a consumer for resources or a predator for prey, V.

K The carrying capacity of the environment, or the maximum density of a given species that can be sustained indefinitely in a particular environment. A high-density equilibrium.

K_N The carrying capacity of species N, usually a prey species.

K_P The carrying capacity of species P, usually a predator or consumer species.

L The relative abundance of a species that competes with the central species for resources; i.e. the density of the other competing species multiplied by its relative ability to compete with the central species.

$\ln$ The natural logarithm of the quantity that follows.

M A community equilibrium. A point in phase space where two or more populations are in equilibrium.

MRT The mean return time. The average time taken for the population trajectory to reach its mean or equilibrium density following a disturbance.

N The density of a population of organisms. In two-species interactions N is the density of a particular resource or prey species.

$\bar{N}$ The mean or average density of a population[32].

N_h The number or density of organisms remaining after harvest.

$N_{h,max}$ The number or density of organisms remaining when harvested at the maximum sustainable yield, Y_{max}.

N_k The number of prey killed by predators.

N_t The density of a population of organisms at time t, usually a particular year.

N_{t-1} The density of a population of organisms at time $t - 1$, usually the previous year.

N_{t-d} The density of a population of organisms at time $t - d$, usually d years ago.

p In economics, the price of a commodity or good. In ecology, the price in energy an organism has to pay to obtain a resource.

PA_d The partial autocorrelation at lag d[32].

PR_d The partial rate correlation at lag d[32].

P The density of a population of consumers or predators.

P_t The density of a population of consumers or predators at time t.

P_{t-1} The density of a population of consumers or predators at time $t - 1$.

P_{t-d} The density of a population of consumers or predators at time $t - d$.

$\Pr(i)$ The probability of the event i occurring.

PRCF   The partial rate correlation function. A histogram of the correlation between the per capita rate of change, R, and the logarithm of population density d time periods previously, where $d = 1, 2, 3, \ldots$

q The encounter rate between predator and prey at which predator searching efficiency for that particular prey reaches its maximum.

Q The coefficient of curvature of the R-function.

Q_j The coefficient of curvature of a low-density R-function (see J).

Q_k The coefficient of curvature of a high-density R-function (see K).

r The correlation coefficient[32]. Also used as the radius of a circle.

r^2 The coefficient of determination[32].

r_d The autocorrelation at lag d[32].

rm The multiple correlation coefficient.

rm^2 The coefficient of multiple determination.

R The realized instantaneous per capita rate of change. The logarithmic rate of change over a unit time period, usually a year, realized by an average organism living in a particular environment and in a population of given density; i.e. $R = \ln N_t - \ln N_{t-1} = \ln(1 + B - D)$.[7]

RT The return time. The time it takes for a population to attain its mean value following a disturbance.

s The standard deviation. The square root of the variance, s^2.

s^2 The variance. The degree of variation in a series of numbers around its mean value[32].

S The total supply of a resource (in ecology) or a good (in economics).

t The time, usually a particular year.

t_h The handling time of a predator, or the time a predator spends in activities other than hunting.

T The total period of time over which predator and prey populations are exposed together.

u	The apparency of a prey species to a predator, or how much attention a predator pays to a particular prey species.
U	The extinction threshold. The density of a population below which it declines automatically to extinction.
v	The attack rate of a predator, or the number of prey killed over a given period of time.
V	The number of prey required by an individual consumer or predator to meet its energetic requirements, or the demand for resources over a given interval of time, usually a year.
VRT	The variance of the return time. The variation in the time taken for the population trajectory to reach its mean or equilibrium density following a disturbance.
w	The energy conversion parameter. Translates energy contained in ingested food into survival and reproduction.
W	The impact of a predator on its prey, or the negative effect of an individual predator on the survival and/or birth rate of a prey individual.
x_t	The error in measuring the value of a variable at time t.
X_i	The value of the ith variable where i is one of several independent variables in a regression model.
X_t	The value of a variable at time t.
Y	The surplus yield. The number of organisms that can be harvested without causing a decrease in the residual population. The net production by a population of size N_h.
Y_{max}	The maximum sustainable yield. The maximum number of organisms that can be harvested in perpetuity without depleting the stock.
Y_i	The value of the ith dependent variable in regression or correlation.
Z	A random normal deviate. A value chosen at random from a normal distribution with mean zero and standard deviation $s = 1$ (see Appendix).

Glossary

Note: terms in *italics* are defined separately.

Absolute population   The total number of organisms in a *population*.

Adapted   Used as an adjective to describe interactions between organisms (usually *competition* or *co-operation*) that are mediated by adapted (evolved) genetic traits; for example, territorial behaviour is an adapted genetic trait that mediates competition for breeding territories between members of the same species.

Aggregation   The tendency for organisms to group together or to have a clumped distribution in space.

Allee effect   Named after the ecologist W. E. Allee[9]. The effect that *incidental co-operation* between organisms has on the *per capita rate of change*, causing it to rise with population density because births either increase or deaths decrease as population density rises.

Asymptotic   A smooth or gradual approach to a constant level, or asymptote.

Attractor   In dynamics, a point or set of points, that attracts dynamic trajectories.

Behavioural response   The attack or feeding response of a predator or consumer to the density of its prey or resource population; specifically the number of prey attacked by a predator in a unit of time expressed as a function of prey density.

Birth rate   The number of offspring produced in a unit of time.

Bottom-up control   The *regulation* or *limitation* of a *population* at or around a stabilizing equilibrium by interactions with resources or food supplies in a lower *trophic level*.

Carrying capacity   The maximum number or density of organisms that can be supported in a given area or environment in perpetuity[18].

Cellular automata   Artificial life forms that occupy cells on a lattice and reproduce and die according to simple rules of interaction with adjacent automata.

Chaos A property of *deterministic* non-linear systems in which oscillations are bounded but do not repeat themselves. Chaotic systems exhibit *sensitive dependence* to their initial conditions[8].

Circular causality A chain of events that leads back to its origin, thereby creating a *feedback loop* in which all components are dependent on each other.

Coefficient of curvature A parameter that affects the curvature of a functional relationship.

Community equilibrium A point where all members of a community of two or more populations are in *equilibrium*.

Competition An interaction or conflict between two or more organisms over resources that are in short supply.

Complex dynamics A term that describes dynamic trajectories that do not converge with time on a stable equilibrium point but exhibit periodic or aperiodic oscillations; for example, *cycles* and *chaos*.

Complex *R*-function An *R-function* that has a complicated shape with usually more than one *equilibrium point*.

Contest competition *Competition* involving contests between two organisms; for example, *territorial behaviour*. See also *adapted*.

Convergent equilibrium An *equilibrium point* that attracts nearby trajectories; an *attractor*.

Cyclic oscillations Oscillations of a dynamic trajectory that repeat themselves at regular intervals. In ecology this usually means that the oscillations repeat themselves every four or more time steps.

Cyrtoid C-shaped. Used in this text in reference to a *behavioural response* of consumers that is shaped like a C.

Damped oscillations Oscillations that overshoot equilibrium but eventually damp out with time to a fixed or stable equilibrium point.

Demand/supply ratio The ratio of the demand for resources by a population of consumers to the supply of those resources.

Density The value of a variable expressed in terms of a unit of space; for example, numbers per square metre.

Density dependence A negative relationship between the reproduction and survival of individuals and the density of the population[12].

Deterministic Completely determined by mathematical equations that contain no *stochastic* effects.

Difference equation A dynamic equation in which the variables change after a discrete interval of time, say a year.

Differential equation A dynamic equation in which the variables change continuously; i.e. the interval of time over which the change occurs approaches zero.

Dimension The number of variables involved in a *feedback loop*.

Diminishing returns Describes a situation in which the output per unit of effort declines as the effort rises. The inefficiency that accumulates as organizations get larger and larger.

Disturbance An unpredictable or random change in an *exogenous* component of the *environment*.

Divergent equilibrium An *equilibrium point* from which nearby trajectories move away or diverge. A *repeller*.

Donor control See *bottom-up control*.

Dynamic Something that changes with time.

Emigration The movement of individuals out of a predefined area or location.

Endogenous Within or part of a mutually causal system. Part of the feedback structure.

Enemy-free space Competition between two or more individuals to avoid attack by predators; for example, competition for a limited number of hiding places or escape routes[13].

Environment All the elements present in an area occupied by a population, including but not limited to food, space, climate, topography, and other organisms of different species.

Environmental forcing Predictable and consistent changes in the environment that affect birth and death rates or the abundance of resources and predators.

Epicentre The place where a pest outbreak starts.

Equilibrium isocline A line in *phase space* along which the rate of change of a variable (or variables) is zero; for example, the *per capita rate of change* of a population is zero on its equilibrium isocline.

Equilibrium point A point in *phase space* at which the rate of change of a variable (or variables) is zero; for example, in population systems, the *per capita rate of change* of the population is zero at an equilibrium point.

Escape threshold An *unstable equilibrium* above which the population escapes from the control of a limiting factor or factors. See also *threshold*.

Exogenous Outside or not part of a *mutually causal* system. A component of the *environment* that affects but is not affected by the population. A *non-reactive* component of the *environment*.

Exponential growth See *geometric growth*.

Extinction threshold An *unstable equilibrium* below which a population declines automatically to extinction.

Feedback dominance Describes the condition where one of many *feedback loops* dominates the dynamics of a system.

Feedback function A function describing the rate of change in a variable in relation to past values of that variable.

Feedback hierarchy Describes the changing sequence of feedback loops that are evoked as population density changes.

Feedback loop A process by which a change in a dynamic variable induces, sometimes through changes in other variables, future changes in that variable. Also called mutual or *circular causality*.

Finite per capita rate of change The *per capita rate of change* over a fixed, or finite, period of time, usually a year.

First order A *feedback loop* that only involves one dynamic variable or a *time lag* of one.

Food chain A series of species that feed one upon the other. See also *trophic level*.

Forcing Predictable and consistent changes in an *exogenous* factor that affect birth and death rates or the abundance of resources and predators.

Function The effect of one variable on another. The statement *A* is a function of *B* means that the value of *A* changes with respect to the value of *B*.

Functional response See *behavioural response*.

Generalist Used to describe a consumer or predator that utilizes several or many different resource or prey species.

Geometric growth A dynamic trajectory defined by a geometric progression through time; for example, 2, 4, 8, 16, 32, ... at time 1, 2, 3, 4, 5, ..., respectively.

Guild A group of different species with the same or similar habits; for example, a group of parasitoids attacking the larval stages of a particular insect.

Higher order Describes a feedback loop that involves two or more dynamic variables.

Hyper-exponential growth See *hyper-geometric growth*.

Hyper-geometric growth A dynamic trajectory that grows or declines faster than a *geometric progression*.

Immigration The movement of individuals into a predefined area or location.

Incidental Used as an adjective to describe interactions between organisms (usually *competition* or *co-operation*) that occur accidentally and do not involve *adapted* (evolved) behaviour; as being in a crowd incidentally decreases ones chance of being eaten by a predator.

Increasing returns Describes a situation in which the output per unit of effort rises as the effort increases. The added efficiency that accompanies co-operative activities as organizations get larger.

Index A relative measure of something. Cannot be related to any standard unit like area or habitat.

Instantaneous Over an instant of time, or as the unit of time over which a dynamic variable is observed approaches zero.

Interspecific Between species. Used as an adjective to describe interactions (usually *competition* or *co-operation*) between different species.

Intraspecific Within a species. Used as an adjective to describe interactions (usually *competition* or *co-operation*) between members of the same species.

Intrinsic A property of a species determined by its innate, genetic characteristics.

Isocline A line joining points of equal value; for example, a zero-growth isocline joins the points where growth is zero, or the population is in equilibrium.

Law of the minimum An empirical "law" stating that growth of a population will be limited by the essential resource that is in shortest supply.

Limitation The prevention of a variable such as population density from attaining a higher value.

Limiting factor An *exogenous* factor that prevents a variable such as population density from attaining a higher value.

Local population A part of a *population* that occupies a particular locality. A population is usually composed of one or more local populations.

Logistic equation The population growth equation first developed by Verhulst as "logistique"[17].

Low-frequency oscillations See *cyclic oscillations*.

Maternal effect An effect that a mother transfers to her offspring[27].

Mean The average of a series of observations. The sum of the observed values divided by the number of observations.

Meta-population A group of *populations* that share occasional *immigrants*.

Negative feedback A *feedback loop* in which the product of the interactions in the loop is negative. The effect of the feedback is to oppose changes in any variable involved in the loop.

Non-reactive Used as an adjective to describe a variable that does not change in response to changes in another variable.

Normal distribution A "bell-shaped" frequency distribution in which the frequency of observations is distributed evenly around the mean.

Numerical response The change in the numbers of consumers (or predators) in response to a change in the numbers of its resources (or prey) due to increased reproduction.

Numerical solution Used when an equation is solved step-by-step, usually by iterating it one step at a time.

Order In dynamic systems, the number of variables involved in the endogenous structure, or the maximum *time lag* in the dynamics. See also *dimension*.

Passive See *non-reactive*.

Per capita Per individual; for example, the per capita *birth rate* is the number of offspring produced per individual, or total number of births over a given interval of time divided by the total population from which those births arose.

Periodic   Describes a phenomenon that repeats itself at regular intervals in time.

Phase portrait   A plot of the trajectory taken by two or more dynamic variables in the *phase space* of those variables.

Phase space   The space defined by a coordinate system composed of two or more dynamic variables.

Population   A group of individuals of the same species that live together in an area of sufficient size that they can carry out their normal functions, including migration, and where emigration and immigration rates are roughly balanced.

Positive feedback   A *feedback loop* in which the product of the interactions in the loop is positive. The effect of the feedback is to amplify changes in any variable involved in the loop.

Predator pit   A depression in the *R-function* of a prey species caused by the *switching* and/or *aggregation* of predators in response to increasing prey density; i.e. the predators have a *sigmoid behavioural response*[10].

Programmed   See *adapted*.

Random   Something that occurs in an unpredictable or unexplainable fashion.

Rate of change   The change in a dynamic variable over a given interval of time.

Ratio dependence   When something depends on the ratio of two things. Usually used to describe situations in which the rate of change of a variable is determined by the ratio of two variables.

Reactive   Used as an adjective to describe a variable that changes in response to changes in another variable.

Realized per capita rate of change   The *rate of change* realized by the average individual living in a particular environment and in a *population* of a given density.

Recipient control   See *top-down* control.

Regulation   The maintenance of a dynamic variable in the vicinity of an *equilibrium point* by a *negative feedback* process.

Repeller   An equilibrium point or set that repels nearby trajectories. A *divergent* equilibrium.

Representative sample   A sample that represents the variability in the population being sampled.

R-function   A function describing how the *realized per capita rate of change* changes with respect to population density.

Satiation   A condition of being full as, for instance, when a consumer has obtained all the resources it needs. See also *saturation*.

Saturation   Describes a situation where a response no longer occurs; for example, the *behavioural response* of a predator to prey density saturates when the prey are so dense that the predator is satiated, in which case adding more prey does not result in more being eaten.

Saw-toothed oscillations   Oscillations that tend to repeat themselves every two time steps and so look like a serrated saw blade.

Scramble competition   *Competition* between two or more organisms that is not mediated by social interactions; i.e. everyone scrambles for what they can get.

Second order   A *feedback loop* that involves two dynamic variables, or a system whose dynamics are influenced by a *time lag* of two.

Sensitive dependence   The current state of a system depends on its initial state. A characteristic of *chaotic* and *transient* systems.

Sensitivity analysis   Determination of the effect of variations in parameter values on the dynamics of a modelled system.

Sigmoid   S-shaped. Used in this text in reference to a *behavioural response* of consumers that is shaped like an S.

Simulation   The *numerical solution* of a model. A time trajectory computed from a dynamic model that simulates the real system.

Stable equilibrium   A *convergent equilibrium* or *attractor*.

Stability   The tendency for a dynamic system to return to or towards a *stabilizing equilibrium* following a disturbance.

Stabilizing equilibrium   An *equilibrium point* to which a dynamic variable is attracted but does not necessarily remain at. The system is maintained near to, but not necessarily at, the equilibrium point.

Standard deviation   The square root of the *variance*[32].

Standard normal deviate   A deviation from the mean of a *normal distribution* which has a *mean* of zero and *standard deviation* of one.

Step-ahead forecasting   Using a discrete model to predict the state of a system one time step into the future.

Stochastic   Involving a random process.

Stratified random sample   A sampling scheme that divides the sampling universe into a number of different strata from which samples are drawn randomly.

Switching   Used to describe the switching of a *generalist* predator from one prey species to another as their relative abundance changes.

System   A group of mutually interdependent objects or variables.

Territorial behaviour   The behaviour associated with obtaining and defending a *territory*.

Territory   An area of habitat occupied by an individual or a group of individuals.

Theory   A systematic statement of the principles, processes and relationships that underlie a particular natural phenomenon.

Threshold   A critical point, usually specifying a change in behaviour. Often used in reference to an *unstable equilibrium* point.

Time delay   The time it takes for a cause and effect to be transmitted around a *feedback loop*.

Time series   A number of observations on the state of a dynamic system taken at equal intervals through time.

Top-down control   The *regulation* or *limitation* of a *population* at or around a *stabilizing equilibrium* by interactions with consumers or predators in a higher *trophic level*.

Transient   A dynamic trajectory that is going somewhere else, usually growing or declining exponentially.

Trophic level   A level in the *food chain*; for example, a plant, herbivore, or carnivore.

Unadapted   An interaction between organisms that does not involve adapted (evolved) traits. See also *incidental*.

Underpopulation   Something that occurs when populations are very sparse.

Unstable equilibrium   See *divergent equilibrium* or *repeller*.

Variance   A measure of the degree of variability around the mean of a series of observations[32].

Vector   An arrow indicating the magnitude and direction of change of a coordinate in *phase space*, usually over a unit of time.

Appendix

TABLE OF STANDARD NORMAL DEVIATES

This table presents a list of random deviates from a normal distribution with a standard deviation of unity; i.e. $s = 1$. To obtain a random normal deviate with a different standard deviation you should multiply the number obtained from the table by the desired standard deviation; i.e. $Z = sX$, where Z is a random normal deviate with standard deviation s and X is a standard normal deviate selected from the table below.

−1.21931	−0.87449	−0.47429	−0.76515	0.31235	0.73104	0.11124	0.86840	−0.03861
−0.46583	2.11258	0.29119	0.77517	1.38159	2.32617	−0.55169	−0.72677	−0.12544
−1.26148	−2.27506	1.87746	0.13395	−0.24859	−1.32934	0.75450	1.22321	0.62659
−1.03658	1.25623	1.47263	−0.28895	−1.48622	0.65377	0.97715	−0.57953	−0.26929
−0.22564	0.73978	0.52595	−0.07190	0.22518	−0.51541	−1.38934	−0.14482	−0.16456
0.25558	−0.28881	0.53236	0.12523	−0.84530	−0.78012	0.83371	0.74061	0.55416
−1.70287	−1.04501	0.38869	1.33952	−0.30850	−1.57078	−0.05302	1.03462	0.01153
0.14151	0.47874	0.50983	1.32368	1.09928	−2.29277	0.63821	−1.05464	0.40511
−0.49635	−0.25102	−0.59470	0.00765	2.02435	−1.25207	−1.73215	0.03076	−0.44995
0.15388	0.48097	−0.27311	1.03804	1.39176	0.64883	−0.15479	−0.20487	0.76096
0.59479	−1.92383	0.99470	0.82672	−0.46702	0.77204	−1.30795	−0.82762	−1.40398
−0.05251	−0.88842	0.65053	1.36570	0.33856	0.10548	0.12362	1.85434	0.50889
−1.21665	1.57323	−0.58759	−1.21186	−0.52527	−0.82395	−2.17079	−0.37513	0.88721
0.73190	0.29634	0.28296	−0.07160	0.00284	0.92224	1.26930	−1.22972	0.05530
−1.51214	−0.68932	1.01042	1.16738	0.04367	0.46431	−0.58050	−0.88954	0.10055
−0.18767	1.85951	0.65215	1.01227	0.64828	−1.75575	−0.36348	1.12469	0.02956
1.16512	2.46261	−1.82073	0.12993	0.92236	−0.27037	−0.20594	0.62772	−0.76302
0.91248	1.09038	−0.49116	−0.36503	1.28823	−0.37213	0.38641	1.11746	−0.48540
0.02766	0.79682	−0.07766	−0.06819	−0.08998	0.26699	0.51679	0.30103	1.05486
−0.36317	−0.82734	0.83918	0.83157	0.22580	0.03460	−0.28468	0.74608	1.33998
−0.18652	−1.04364	0.36618	0.85801	−1.47218	0.90907	0.43936	−0.59768	0.97826
0.41502	−0.30580	−0.26628	0.37283	1.36598	1.06134	−0.76690	1.01617	0.28585

−0.02116	1.22607	−1.09249	1.03595	1.80189	−0.15630	0.24300	0.87451	1.47278
−1.60613	−0.33846	0.42676	1.05506	0.25272	−1.38413	0.15274	1.26128	0.38356
−0.41161	1.05177	0.74470	−0.49831	−0.56643	−1.13986	−0.51807	−1.08643	−0.79203
0.80614	−1.75873	0.99838	1.90137	−0.55202	−0.16590	0.00716	−0.00594	0.70516
−0.74629	−0.30681	0.97672	0.54890	−0.87386	−3.86193	−0.18161	0.12709	−0.16017
−1.10469	1.14282	−1.36047	0.77862	−2.11435	−0.12915	0.88671	0.09429	2.05503
−1.86970	0.34695	1.23869	0.88750	1.19431	−0.65036	−1.86514	0.26245	0.39178
−1.25089	1.47562	0.53957	0.65319	−0.16231	0.44841	0.67732	−0.90396	1.08222
1.92812	1.03643	−1.53050	−1.60716	−0.58092	−0.24806	0.85821	−0.91535	1.65170
0.92036	1.21339	1.52106	0.78111	−0.10426	1.08821	−0.51035	0.24011	0.21557
−0.71865	−0.28391	1.53200	−0.89525	−0.09025	−0.55571	−0.00334	−1.03907	0.90809
−0.17098	2.55476	0.53896	1.94829	0.15045	−0.20731	−1.42544	0.57042	1.16027
0.08718	0.86317	0.16845	−0.75221	1.42424	0.39767	1.20138	0.91979	−0.99776
0.87597	−0.42962	1.25869	−0.27245	0.58717	−0.36270	0.58795	−0.65455	−1.07391
−0.15551	1.38652	0.98079	0.88419	−0.51842	−0.51018	−1.52564	−1.11589	−1.38748
−0.52380	0.61770	−0.63861	−0.32656	0.61472	−1.69211	0.27052	0.48271	−0.53602
0.93329	0.87023	0.77648	1.38916	−0.47823	−0.70378	2.27889	−0.62352	−0.50294
−1.03749	−0.10657	0.68955	2.22589	−0.57237	1.12929	3.04394	0.82639	0.15954
0.48717	−0.34131	0.81531	−0.50465	−0.43274	−0.05854	0.88369	−0.76679	−0.31511
0.04242	−1.52988	−2.32176	−1.89529	−0.36101	−1.16877	−1.05861	0.75576	0.36827
−0.43544	−0.55214	1.37509	−0.31118	−0.73713	−1.39318	1.47133	−0.24679	0.76036
0.26756	0.78183	0.37551	1.41905	0.07372	−0.33265	0.36761	1.13099	0.02972
−1.56935	−0.64766	−0.89739	1.10502	0.28989	2.12856	0.23570	0.61305	0.00630
1.14439	−1.58346	−1.24826	−0.60979	−1.15395	−0.70749	0.28311	1.32557	1.54883
−1.78596	−1.36182	−0.18057	−0.13205	−0.53711	−1.08763	0.53859	−1.15431	0.02911
−0.26089	0.35671	0.41110	0.29425	0.61318	1.52479	0.70828	1.19528	0.18976
−0.22427	0.70349	1.17590	0.78900	−1.64214	−0.55865	−0.52625	−0.70142	0.24625
−0.47436	1.18541	0.05680	−1.29883	0.59470	0.67479	0.96274	−0.89891	−2.48170
−0.25661	−0.54763	−0.52641	−1.73233	−1.63563	0.14765	0.66609	−0.80810	0.52282
0.71245	−0.13966	0.11948	0.19287	−0.85959	1.19503	0.84633	1.23188	−0.12704
1.55733	0.52545	1.11142	0.93284	0.78452	0.57034	−0.97032	−1.99237	−0.01568
−1.01320	1.00643	−1.26815	0.29425	0.36143	−0.12741	−0.46572	0.10324	−0.01435
0.01402	−0.54719	0.61075	−0.48000	1.36900	1.38964	−0.24081	−0.52558	0.49851
1.71296	−0.32123	0.27387	0.43336	−0.05284	0.47202	0.57721	−0.52490	0.79849
−0.48053	0.73552	−0.15783	0.76984	−0.57457	0.26418	−1.52786	−0.84985	−0.58362
−1.93720	−1.46045	−0.48733	2.18511	0.43184	−0.91335	0.09849	1.51035	0.25939
−0.54310	−0.11521	1.02985	−0.50972	−0.68755	−2.13173	1.46431	2.31206	−1.00505
0.35326	−1.75344	−1.33640	1.48236	1.20082	0.60292	0.85216	1.54327	−0.53599
0.70379	−1.43772	1.83021	−1.10028	−0.30906	0.43809	−1.21702	1.26788	−0.05320
−1.74864	−0.45418	−1.03768	0.21324	−1.01764	0.39409	−0.49264	−0.26227	0.11456
−0.23616	0.34996	1.92982	1.02280	0.62990	1.20340	0.22823	−0.87905	0.49761
−0.24567	0.50784	−0.85205	0.70875	2.55901	0.61621	−1.13813	−1.11762	0.90114
−0.20230	0.44724	0.38950	−0.24810	−0.73791	1.05994	−0.24082	0.42758	−0.58136
0.53351	−0.49190	1.27310	−0.10641	0.47940	0.50187	−0.53179	−1.95383	1.28945

−0.31968	−1.60643	0.49872	0.76985	0.24557	1.20560	0.16185	−2.64521	0.45169
−0.13502	1.34062	−0.22250	0.33976	−0.89081	0.71260	1.41375	−0.35309	−0.72616
−0.49410	1.35955	−0.25882	−0.21113	−0.45056	1.49801	1.28382	1.15539	0.86265
−1.82949	0.70164	−1.51953	−0.52677	0.17153	1.19122	0.81665	0.35929	−1.14709
−0.11070	0.09243	−0.32516	−0.02837	0.33443	−1.49042	1.42427	−0.78860	−1.44093
0.09535	0.65885	−0.63499	0.26247	0.07619	1.07059	0.98512	1.72468	−0.11234
0.14442	−0.28779	2.49992	−1.14004	−0.73631	0.07364	1.47726	0.87426	−0.08642
0.36878	−0.45661	−0.22312	0.34877	2.06458	1.16247	−0.60297	0.06417	1.78317

Index

Milton Keynes UK
Ingram Content Group UK Ltd.
UKHW040105071024
449327UK00019B/837